ENERGY SCIENCE, ENGINEERING AND TECHNOLOGY

THE LONG ISLAND SOLAR FARM

A TRAILBLAZING RESOURCE FOR DEVELOPMENT AND PARTNERSHIPS

ENERGY SCIENCE, ENGINEERING AND TECHNOLOGY

Additional books in this series can be found on Nova's website under the Series tab.

Additional E-books in this series can be found on Nova's website under the E-book tab.

ENERGY SCIENCE, ENGINEERING AND TECHNOLOGY

THE LONG ISLAND SOLAR FARM

A TRAILBLAZING RESOURCE FOR DEVELOPMENT AND PARTNERSHIPS

CLARA MCKENZIE
EDITOR

nova publishers
New York

For permission to use material from this book please contact us:
Telephone 631-231-7269; Fax 631-231-8175
Web Site: http://www.novapublishers.com

Additional color graphics may be available in the e-book version of this book.

Library of Congress Cataloging-in-Publication Data

ISBN: 978-1-63117-419-3

Published by Nova Science Publishers, Inc. † New York

CONTENTS

PREFACE

The Long Island Solar Farm project came together as a joint effort of five very different interest groups: a federal agency, a research institution, an electric utility, a private business, and the general public. The project is remarkable for three major reasons: first, it is the largest utility-scale solar power plant in the Eastern United States; second, it is a commercial project built on federally administered public lands; and third, the project was very unlikely to have started in the first place. The process by which the Long Island Solar Farm was developed is intricate and unusual. This reflects many of the nuanced conditions that made siting the solar farm on federal property a unique opportunity for very different groups. This book discusses project concepts; the process involved in executing the projects; and also provides an environmental assessment for the BP solar array project from the Brookhaven National Laboratory.

Chapter 1 – In November 2011, a utility-scale solar array became operational in the most unlikely of places, at Brookhaven National Laboratory in Eastern Long Island, New York. The Long Island Solar Farm project came together as a joint effort of five very different interest groups: a federal agency, a research institution, an electric utility, a private business, and the general public. The project is remarkable for three major reasons: first, it is the largest utility-scale solar power plant in the Eastern United States; second, it is a commercial project built on federally administered public lands; and third, the project was very unlikely to have started in the first place.

The process by which the Long Island Solar Farm was developed is intricate and unusual. This reflects many of the nuanced conditions that made siting the solar farm on federal property a unique opportunity for very different groups. Though many of these nuances make the Long Island Solar Farm difficult to imitate as a template, the research aspects of the project make it a trailblazing resource to inform future development of photovoltaic (PV) solar projects in the East. Furthermore, the innovation in attitude it took to develop this project serves as an excellent model for large-scale solar power development and public-private partnerships in general.

Chapter 2 – BNL is a national laboratory overseen and primarily funded by the Office of Science (SC) of the DOE, and operated and managed by Brookhaven Science Associates (BSA), a limited liability company founded by Stony Brook University and Battelle, a nonprofit, applied science and technology organization. Located 60 miles east of New York City in Upton, NY, BNL conducts research in high energy and nuclear physics, chemistry, nanotechnology, environmental sciences, energy technologies and national security. Among its missions, the Laboratory is charged with conceiving, designing, constructing and operating world-class, complex, leading-edge research facilities in response to the mission needs of DOE and to a large community consisting of university, industry, government and international users [SER].

If constructed, the proposed BP Solar Project would be located on approximately 200 acres and would encompass two large-scale commercial solar photovoltaic (PV) arrays totaling approximately 37 MW, and potentially a smaller Laboratory dedicated array of 1 to 2 MW, with an associated support building. This project grew out of solicitation by the Long Island Power Authority (LIPA) to diversify its portfolio with renewable energy sources. LIPA publically solicited companies for proposals to provide solar-generated power on Long Island. In that solicitation, DOE offered BNL property as a host site for a project that would meet LIPA's needs. BP Solar's successful proposal to LIPA for a 37 MW photovoltaic system on the federal property resulted in negotiations between DOE and BP Solar for access to the site for construction and operation of the system. The project supports DOE's renewable portfolio and the Department's vital goals of reducing dependence on fossil fuels and encouraging research and development in energy technologies. Additionally, this proposed solar project is consistent with NY State's goal to meet 45 percent of its electricity needs through improved energy efficiency and renewable sources by the year 2015.

In: The Long Island Solar Farm
Editor: Clara Mckenzie
ISBN: 978-1-63117-419-3

Chapter 1

THE LONG ISLAND SOLAR FARM*

Robert S. Anders

ABSTRACT

In November 2011, a utility-scale solar array became operational in the most unlikely of places, at Brookhaven National Laboratory in Eastern Long Island, New York. The Long Island Solar Farm project came together as a joint effort of five very different interest groups: a federal agency, a research institution, an electric utility, a private business, and the general public. The project is remarkable for three major reasons: first, it is the largest utility-scale solar power plant in the Eastern United States; second, it is a commercial project built on federally administered public lands; and third, the project was very unlikely to have started in the first place.

The process by which the Long Island Solar Farm was developed is intricate and unusual. This reflects many of the nuanced conditions that made siting the solar farm on federal property a unique opportunity for very different groups. Though many of these nuances make the Long Island Solar Farm difficult to imitate as a template, the research aspects of the project make it a trailblazing resource to inform future development of photovoltaic (PV) solar projects in the East. Furthermore, the innovation in attitude it took to develop this project serves as an

* This is an edited, reformatted and augmented version of Technical Report DOE/GO-102013-3914, issued by Brookhaven National Laboratory, U.S. Department of Energy, dated May 2013.

excellent model for large-scale solar power development and public-private partnerships in general.

Executive Summary

The Long Island Solar Farm (LISF) is a remarkable success story, whereby very different interest groups found a way to capitalize on unusual circumstances to develop a mutually beneficial source of renewable energy. The uniqueness of the circumstances that were necessary to develop the Long Island Solar Farm make it very difficult to replicate. The project is, however, an unparalleled resource for solar energy research, which will greatly inform large- scale PV solar development in the East. Lastly, the LISF is a superb model for the process by which the project developed and the innovation and leadership shown by the different players.

Unusual Circumstances

The Long Island Solar Farm exists at a nexus of very favorable conditions that were each crucially important to its development. The project began with what some might assess as an unnecessarily bold request for proposals from the electric utility, which was looking for a very large amount of PV solar power. That, in turn, opened the door for companies to consider developing a single-source utility-scale array on densely-populated Long Island, where the vast majority of existing PV solar comes from very small distributed arrays. Due to very limited land constraints on the island, BP Solar looked at the possibility of building a project on federal lands at Brookhaven National Laboratory (BNL). All parties involved were quick to recognize that the proposed site was replete with fortunate circumstances:

- No transmission was necessary; an existing substation was immediately next to the site
- Availability of large tracts of unused land on federal property in close proximity to load
- BNL is a "Superfund site," which produced an existing body of thorough environmental data
- BNL's history as a military base created a trove of historical and cultural information

- The federal facility's mission was such that a commercial project could be made mutually beneficial.

An Unlikely Project

The Long Island Solar Farm was, by all accounts, a tremendous leap outside the comfort zone for each of the participating groups: a federal agency (U.S. Department of Energy), a research institution (Brookhaven National Laboratory), an electric utility (Long Island Power Authority), a private business (BP Solar), and the general public. Each of these groups could have easily denied the project or refused to participate on the grounds that such a project was not what their organization "usually does" or it would "set a wrong precedent." Had any single group decided not to participate, or wholly object to it—reflecting their most likely attitude—then this project would have never taken place. Instead, and despite the odds, each of these very different interest groups found a way to make the project succeed.

Figure A-1. Arial View of the Long Island Solar Farm at Brookhaven National Lab.

Innovation in Attitude

Each of the groups exhibited excellent leadership within their respective organizations. No single dominant leader emerged in the consortium, which fostered a truly collaborative effort. Each group faced significant risks in this endeavor, but found innovative ways to safeguard their interests and mitigate those risks. The hallmark agreements of the project are: the use of an easement versus a lease for the land, the use of consideration as a means of compensation, the development of a dedicated research array, and the creation of a robust natural resources benefits package to mitigate environmental impacts.

A Trailblazing Resource

The conditional nuances of the Long Island Solar Farm make the development of a similar PV solar power plant very unlikely. This project does, however, serve as a very important data center, whereby enormous amounts of information about the array are being collected for analysis and research. The information gleaned from the solar farm and the dedicated research array will spur innovations in meteorological forecasting, PV solar technology, component systems design, large-volume storage, and utility-scale grid integration, among other areas of research. Though it may be difficult to build another Long Island Solar Farm, the lessons learned from its exposure to the dynamic conditions on the Atlantic coast certainly make it an invaluable resource to help inform future development of PV solar power in the Eastern United States.

A Model for Development

The organizational schematic in Figure ES-1 shows the dynamic results of the collaborative effort. Unsurprisingly, there were a great deal of negotiations behind each of these interactions and agreements. Each group had a sense for the project's potential, though, which kept them committed to its success. The outcome was the successful accomplishment of a very ambitious project that has already generated significant benefits for each of the parties involved. It is this process by which the Long Island Solar Farm was conceived, negotiated, and developed that make it an excellent example of public-private partnerships in general.

INTRODUCTION

Generating power by converting sunlight into electricity is not a new concept. Neither is generating solar power at the utility scale. Photovoltaic (PV) panels have been collecting photons for decades, and PV arrays have been generating power for utilities since the first megawatt- scale solar farm was built in Sacramento, California, in 1984 (Green Energy News 2009).

What is new, however, is the accelerating demand for clean energy, particularly PV solar energy. The modern surge for solar is, in part, driven by rising demand for electricity and increasing environmental costs associated with conventional fuels. In recent years, large-scale solar energy development has also been invigorated by the economic forces of technological innovation, falling costs of production, and political support in the way of renewable energy standards and goals.

As a result, numerous large-scale solar projects have taken root domestically and internationally, and they are continuing to grow. Understandably, many of the earliest projects were developed where the sun shines the most. In the United States, for example, that has been primarily in the western portion of the country, more specifically, in the Southwest. The Western United States is certainly attractive for solar development for many reasons: land tends to be cheap, environmental impacts tend to be less complex, the population is comparatively less dense, the solar irradiance is high, humidity is low, and the weather is predictably cloudless for most of the year.

Though the conditions out West are rather ideal, large-scale solar power is still very much a viable source of renewable energy in a myriad of conditions and locations. In the Eastern United States, particularly in the Northeast, for instance, the conditions are dramatically different from the Southwest—land tends to be expensive, environmental impacts tend to be very complex, the population is more dense, the solar irradiance is comparatively less, humidity can soar, and the weather is highly variable and extremely difficult to predict. Even though the conditions may not be as ideal as those found in the West, economic forces are spurring the feasibility of PV solar power development in the East.

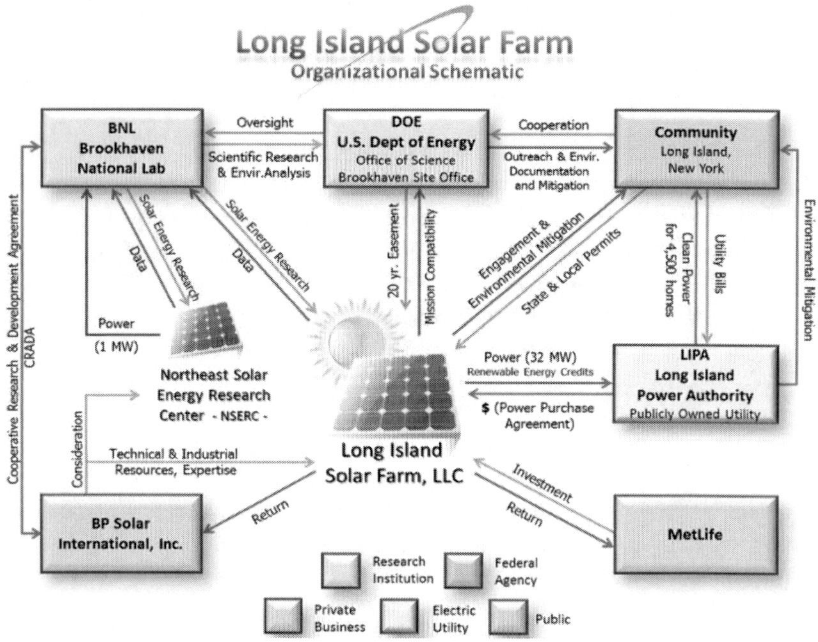

Figure ES-1. Organization of the Long Island Farm Participants.

The Eastern United States has thus become a new frontier for utility-scale PV-solar energy development. Accordingly, there are few projects substantial enough to demonstrate PV solar energy's viability as a large-scale renewable resource for utilities in the East.

The Long Island Solar Farm (LISF) has strong potential to serve as that example. The LISF is a mere 60 miles from downtown New York City and is the largest solar array east of the Mississippi River. At the time of its development, it was ranked 5th among the largest PV power plants in the United States (and tied for 25th in the world) in terms of capacity (Lenardic 2012).

Though there are many lessons learned in the conception and development of the Long Island Solar Farm, it is best viewed as an example of the bold innovation in attitude it took to create a mutually beneficial source of clean, renewable power in what might otherwise seem an unlikely place for a large-scale PV solar farm.

PROJECT PARTICIPANTS

• Federal Agency	U.S. Department of Energy
• Research Institution	Brookhaven Science Associates, BNL
• Private Industry	BP Solar, Inc.
• Electric Utility	Long Island Power Authority
• Environment	The Long Island Pine Barrens Society

U.S. Department of Energy, Office of Science, Brookhaven Site Office

The mission of the U.S. Department of Energy (DOE) is to ensure America's security and prosperity by addressing its energy, environmental, and nuclear challenges through transformative science and technology solutions. The department's strategic goals to achieve the mission are designed to deliver results along four strategic themes: Transformation of the Nation's Energy System; The Science and Engineering Enterprise; National Security; and Management and Operational Excellence.

The mission of the Office of Science is to foster, formulate, and support forefront basic and applied research programs that advance the science and technology foundations necessary to accomplish DOE missions. The Office of Science is responsible for six multi-program national labs across the country and is the single largest supporter of basic research in the physical sciences in the United States.

The DOE Brookhaven Site Office is located at Brookhaven National Laboratory in Upton, New York. The mission of the Brookhaven Site Office is to manage and administer DOE's performance-based contract with the Brookhaven Science Associates for the management and operation of Brookhaven National Laboratory.

The Brookhaven Site Office supports the Office of Science mission to deliver the remarkable discoveries and scientific tools that transform our understanding of energy and matter, and advance the national, economic, and energy security of the United States (Office of Science website, accessed August 2012).

Henceforth in this document, "DOE" will be used to refer to the Department of Energy's Office of Science, Brookhaven Site Office at Brookhaven National Laboratory.

DOE's principals in this project are:

- Robert J. Gordon, Director of Business Management
- Louis Sadler, Chief Counsel

Brookhaven Science Associates, Brookhaven National Laboratory

Brookhaven Science Associates (BSA) was established for the sole purpose of managing and operating Brookhaven National Laboratory.

BROOKHAVEN
NATIONAL LABORATORY

Formed as a 50/50 partnership between Battelle and the Research Foundation of State University of New York (SUNY) on behalf of Stony Brook University (SBU), BSA is the legal entity responsible for leading BNL successfully through the 21st century.

Both SBU and Battelle are strongly motivated to ensure Brookhaven's success. Being Brookhaven's closest university neighbor, SBU is the single largest user of BNL facilities; BNL facilities and scientific staff are essential to the vitality of the university's intellectual life and to the impact of many of its research programs.

BSA is governed by a 16-member board of directors: five appointed by the Research Foundation, five by Battelle, and one each from six of the nation's premier research universities (Columbia, Yale, Cornell, Harvard, Princeton, and the Massachusetts Institute of Technology).

Henceforth in this document, "BNL" or the "lab" will refer to the lab site itself or the institution. "BSA" refers to the Brookhaven Science Associates staff, whose principals in this project were:

- J. Patrick Looney, Ph.D., assistant laboratory director for policy and strategic planning, Sustainable Energy Technologies Department
- Robert J. Lofaro, group leader, Renewable Energy Group, Sustainable Energy Technologies Department
- Timothy Green, Ph.D. , manager, Environmental Protection Division

BP Solar

British Petroleum Solar International, Inc., commonly referred to as "BP Solar," is a former subsidiary of British Petroleum. It was a manufacturer and installer of photovoltaic solar cells headquartered in Madrid, Spain, with production facilities in India and the People's Republic of China. Operating since 1981, the company became wholly owned by BP in the mid-1980s and was closed on Dec. 21, 2011 (Bergin 2011).

BP Solar was selected from a competitive process after Long Island Power Authority (LIPA) published a request for proposals for PV solar development in May 2008. As of this publication, BP is still under contract with DOE and BSA regarding the Long Island Solar Farm and its complementary research array. BP Solar is working to sell their easement rights and transfer obligations to a third party.

BP Solar's principal in this project was Richard Chandler, Project Manager

Long Island Power Authority (LIPA)

In May 1998, the LIPA became Long Island's primary electric service provider. LIPA is a non-profit municipal electric provider, owns the retail electric transmission and distribution system on Long Island, and provides electric service to more than 1.1 million customers in Nassau and Suffolk Counties and the Rockaway Peninsula in Queens. LIPA is the 2nd largest municipal electric utility in the nation in terms of electric revenue, 3rd largest in terms of customers served, and the 7th largest in terms of electricity delivered. In 2010, LIPA outperformed all other overhead electric utilities in

New York state for reduced frequency of service interruptions and ranked 2nd for shortest duration of service interruptions. LIPA does not provide natural gas service or own any on-island generating assets (LIPA website, August 2012).

LIPA
Long Island Power Authority

LIPA's principal in this project was Michael J. Deering, Vice President of Environmental Affairs.

The Long Island Pine Barrens Society (LIPBS)

The Long Island Pine Barrens Society is an environmental education and advocacy organization focusing on protecting drinking water and preserving open space, especially in Long Island's Pine Barrens.

LONG ISLAND
PINE BARRENS
SOCIETY

The LIPBS is a nonpartisan, not-for-profit organization dedicated to the study, appreciation, and protection of these unique woodlands. Founded in 1977, the society has become one of Long Island's most effective champions of preserving natural resources through sound land use. The society influences the public debates regarding land use, conservation, and other environmental issues through its scientific research and programs of public education and advocacy.

The LIPBS should not be confused with the Central Pine Barrens Joint Planning & Policy Commission, commonly referred to as the "Pine Barrens Commission." The Commission was established by the New York State Long Island Pine Barrens Protection Act of 1993. It has jurisdictional legal authority over the land use in the Central Pine Barrens region. The Pine Barrens Society, by contrast, has no jurisdictional authority.

The LIPBS was the largest voice of opposition to the development of the Long Island Solar Farm. Its principal is Richard Amper, Executive Director.

The Central Pine Barrens Joint Planning and Policy Commission

In 1993, New York State's Long Island Pine Barrens Protection Act defined this region at the junction of the towns of Brookhaven, Riverhead, and Southampton. The 1993 act created a five- member Central Pine Barrens Joint Planning & Policy Commission and an advisory committee, and mandated the production and implementation of the Central Pine Barrens Comprehensive Land Use Plan, adopted in June 1995. Under New York Environmental Conservation Law Article 57, the commission is responsible to produce, implement, and administer a Comprehensive Land Use Plan. The act and the plan charge the commission with the combined duties of a state agency, a planning board, and a park commission:

- Land use review, permitting, and enforcement authority in the Central Pine Barrens, along with the local municipalities
- Establishment and operation of a transferable development rights and conservation easement program
- Coordination of public lands stewardship and management on a regional basis.

The Long Island Solar Farm (LISF) at a Glance

The LISF is a 32-megawatt (MW) solar photovoltaic power plant built through a collaboration that included BP Solar, LIPA, and DOE. The LISF, located on the Brookhaven National Laboratory site, began delivering power to the LIPA grid in November 2011 and is currently the largest solar photovoltaic power plant in the Eastern United States. It is generating enough renewable energy to power approximately 4,500 homes, and is helping New York state meet its clean energy and carbon reduction goals (BNL, August, 2012).

Project Location	• Brookhaven National Laboratory, Upton, New York
Land Area	• Approximately 200 acres
Property Type	• Federal land (administered by the DOE)
Host	• DOE issued a 20-year easement to LISF, LLC to use land at • Brookhaven National Laboratory
Developer/ Owner	• LISF, LLC is jointly owned by BP Solar and MetLife, Inc.
Workforce	• 200+ FTE during construction, two during operation
Operator	• The day-to-day operation and maintenance of the LISF are contracted to True South, Inc.
Technical Design	• 164,312 ground-mounted panels (24 panels / rack) • 180° azimuth • Fixed-tilt at 27° (for optimized capacity during summer months) • 25 power blocks with dual inverters
Special Design Feature– Dedicated Research Array	• The 32-MW LISF was built in response to a request for proposals (RFP) issued by LIPA. Part of the agreement for use of the federal property included the development of a dedicated research array, called the Northeast Solar Energy Research Center (NSERC). The LISF generates 32 MW of commercial power. The 1 MW generated by the NSERC is retained and used on site at Brookhaven National Laboratory.
Technology	• Polycrystalline silicon solar PV modules
Commissioned	• Nov. 3, 2011
Lifespan	• Estimated to be 40 years (2051)
Peak Capacity	• 32 MW (AC)
Annual Energy Output	• 44 gigawatt (GW)/hours (estimated annual average), equivalent to the annual usage of approximately 4,500 homes
Utility	• LIPA purchases 100% of the LISF output, according to a 20-year power purchase agreement it has with LISF, LLC (expires November 2031)
Power Distribution	• To LIPA residential and commercial customers in Eastern Long Island

Map

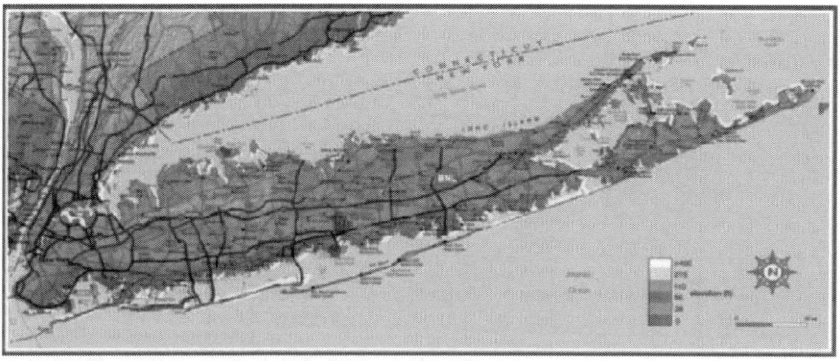

Figure 1. Location of Brookhaven National Laboratory.

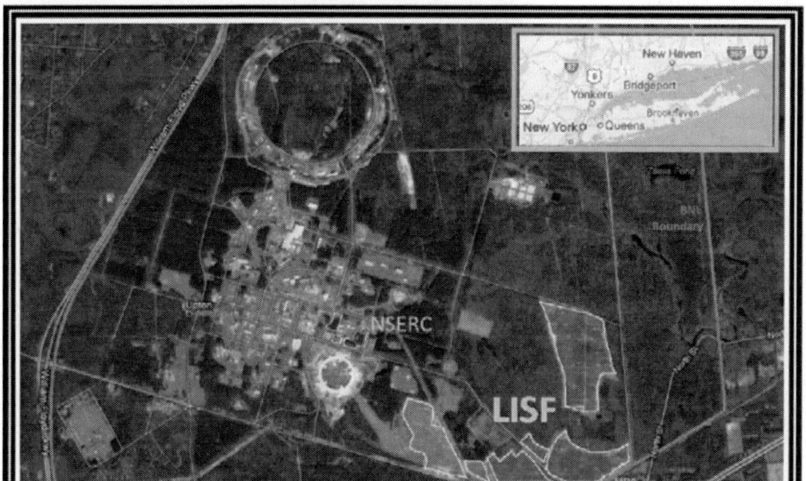

Figure 2. Satellite View of Brookhaven National Laboratory.

HISTORICAL AND CONTEMPORARY CONTEXT

In order to understand the development of the Long Island Solar Farm and to appreciate its size and scale, it is important to consider the historical and contemporary context of the project. This section provides a broad overview of the Long Island Solar Farm as it relates to PV solar development in the rest of the world and across the United States. It also establishes a backdrop of the state and local conditions that helped shape the development of the project.

Throughout the World

The use of photovoltaic power generation has been accelerating in the United States for the last decade. This is a trend that mirrors the development of PV systems around the globe, which has also been rapidly expanding. As a share of the world's installed PV capacity, the United States is ranked 4th (tied with Spain), behind Germany, Italy, and Japan (EPIA, 2012).

Naturally, the size of the individual PV systems varies dramatically throughout these different parts of the world. Countries with lesser overall installed capacities tend to rely on small-scale distributed generation. More ambitious countries, on the other hand, have developed very large utility-scale arrays in conjunction with their distributed programs. In this context on the world stage, the Long Island Solar Farm is tied for 25th in terms of power output.[1] Considering the thousands of generation facilities that have been built around the globe, the LISF is a remarkably large array.

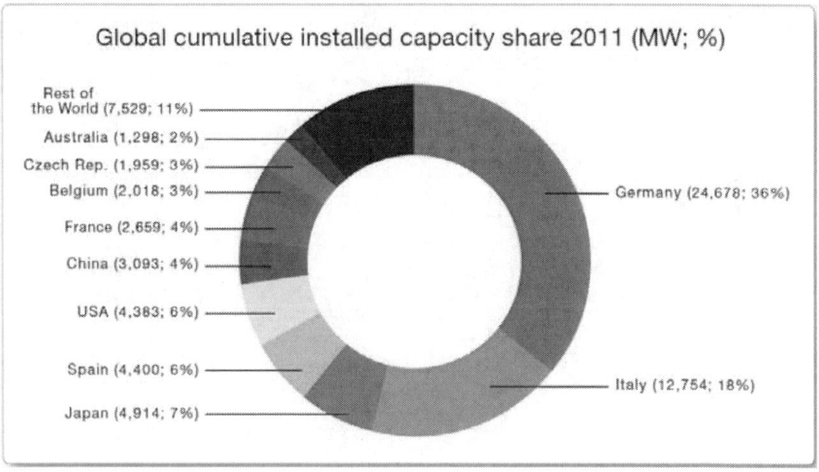

Figure 3. Global cumulative installed capacity share 2011.

Across the United States

On the domestic front, the Long Island Solar Farm establishes itself among the largest PV solar arrays in the country. At the time of its commissioning, the LISF was the 5th largest PV plant in the United States and by far the largest facility in the East.[2]

One of the most notable distinctions in Table 1 is that the Long Island Solar Farm is the only PV facility of its size in the Northeast. It comes as no surprise that the largest arrays are concentrated in the Western states. The solar resource map of the United States clearly identifies the Southwest as the region with the most intense and consistent irradiance, which produces the best solar energy yield. This, among other factors, makes large- scale PV projects very attractive in the West.

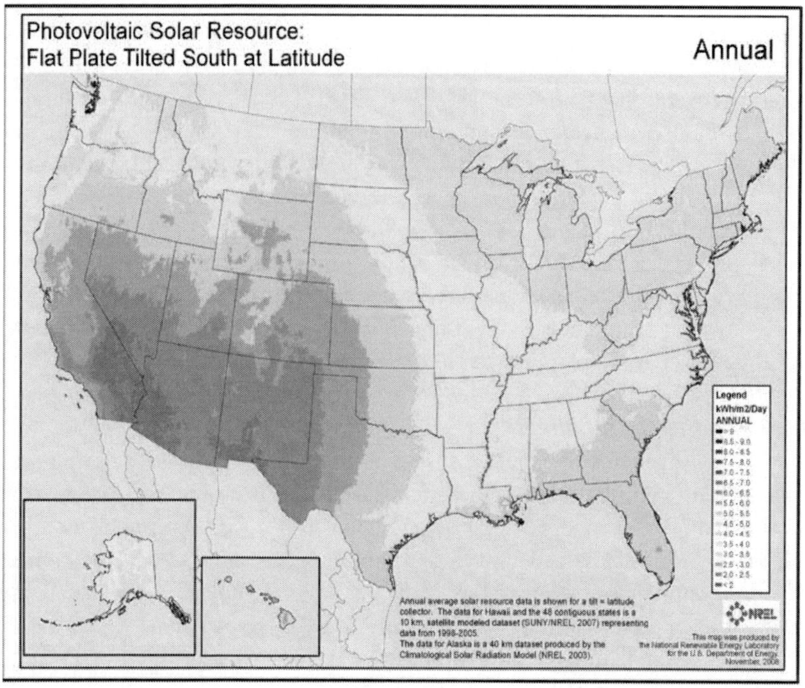

Figure 4. Photovoltaic solar resource: flat plate tilted south at latitude.

Table 1. PV Solar Power Generation Facilities in the United States

Rank	Power	Country	Location	Description	Commissioned
1	48 MW	USA	Boulder City, NV	Copper Mountain Solar Facility	2010
2	45 MW	USA	Kettleman Hills, CA	Avenal Solar Facility	2011
3	42 MW	USA	Sonoran desert, AZ	Mesquite Solar I	2011
4	35 MW	USA	Mosca, CO	San Luis Valley Solar Ranch	2011
5	32 MW	USA	Long Island, NY	Long Island Solar Farm	2011
6	30 MW	USA	Colfax Co, NM	Cimarron Solar Facility	2010
7	30 MW	USA	Webberville, TX	Webberville Solar Farm	2011
8	25 MW	USA	Arcadia, FL	DeSoto Next Gen Solar Energy Center	2009

Policy and Law

There are, of course, other forces driving utility-scale solar development. A myriad of political influences, federal and state policies, and technological reasons are making large-scale solar more attractive in places where the conditions are less than ideal. For starters, the Energy Policy Act (EPACT) of 2005 was a pivotal piece of federal energy legislation—the first comprehensive energy law in 13 years. EPACT authorizes and encourages, through the use of incentives and grants, energy production, conservation, and improved efficiency in permitting new facilities. It also helps set conditions to spur investments in renewable energy development through the use of tax credits and other incentives (LIPA 2004). Federal tax credits, like the production tax credit (PTC), helped catalyze investment and development in a broad sense. A provision of the PTC expired at the end of 2011, which caused a rush to get projects, like the Long Island Solar Farm, online before that deadline for credit application expired.

Federal Agency Initiatives

Federal agencies have taken strides to unlock the potential of large-scale solar opportunities around the country. The U.S. Department of the Interior, for example, has developed solar development zones (SDZ) on public lands in the West. SDZs are selected areas that minimize the impacts of development. The environmental characteristics of SDZs have already been (or are being) inventoried so as to facilitate shorter permitting timelines.

Other federal initiatives have been made by the Department of Defense. The DOD is the country's single largest consumer of energy and spends $4 billion every year on facility energy consumption (SERDP Website). In 2002, Congress appropriated funds for the DOD to do a comprehensive energy assessment, which it completed in 2005 (Snook 2007). Subsequently, through programs like the Strategic Environmental Research and Development

Program (SERDP) and the Environmental Security Technology Certification Program (ESTCP), the DOD is investing in innovative technologies, including PV solar, to help it achieve its renewable energy goals.[3] The programs use federal installations as test beds for cutting edge technologies developed by researchers from industry, universities, and federal agencies (SERDP 2012).

In concert with the DOD, DOE is looking at its facilities with an eye for renewable energy resourcing. In 2009, DOE created an Energy Parks Initiative (EPI) to take a focused look at using existing landfills, brownfields, greenfields, and federal facilities as opportunities for renewable energy development. EPI is designed to foster effective partnering of DOE, other federal agencies, private industry, state and local governments, local and regional communities, and stakeholders (Gilbertson 2009).

In the last decade at the national level, actions by Congress and initiatives taken by major federal agencies, such as the U.S. Department of Interior, DOD, and DOE, have significantly helped drive innovation and expansion of PV solar development throughout the country.

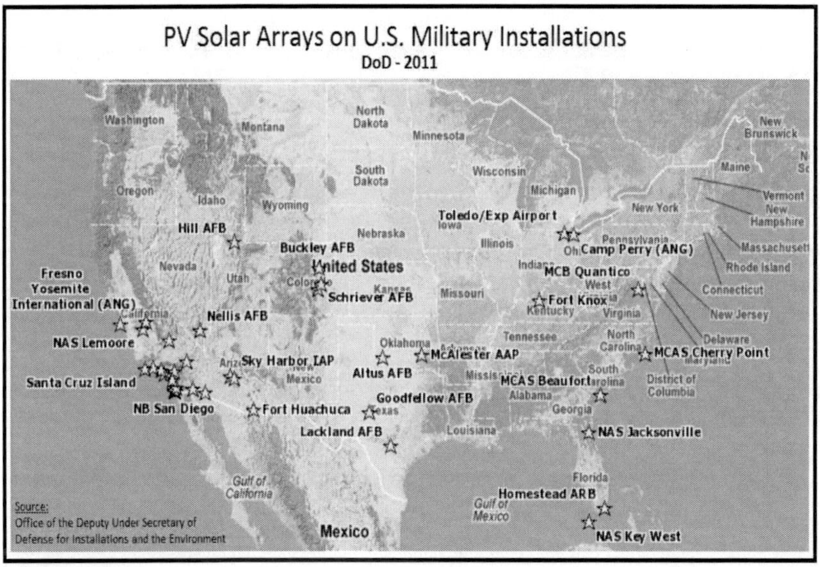

Figure 5. PV solar arrays on U.S. military installations.

Among the States

At the state level, many legislatures have adopted or are considering the use of renewable portfolio standards (RPS). These RPSs are designed to steer electricity resource development away from fossil energy toward cleaner sources of power.

Figure 6 (FERC 2011) shows an uneven distribution of RPSs across the country, with varying degrees of intensity, both in terms of percentage goals and time horizons. Those states that have adopted RPSs, or similar programs, are generally those that have large and increasing load projections, and a comparative scarcity of power resources. The Southeast, in contrast, has thinner populations and access to very cheap power. As a result, those states are less inclined to expand renewable resources to higher proportions of their energy mix.

In New York

New York state—and the Northeast generally—is subject to very heavy and increasing loads associated with dense populations and urban development. The region also relies heavily on conventional fuel resources, which are volatile and costly. In confronting these conditions, New York has emerged as one of the most aggressive states in the country with regard to the breadth and boldness of its policies toward renewable energy development. For example, New York's RPS—to generate 30% of its electricity from renewable resources by 2015[4]—establishes the highest percentage of renewable power deployment in the country, matched only by California and Colorado (FERC 2011).

New York has also established clean energy initiatives, green power purchasing programs, and various other polices to encourage residential and commercial development of PV solar (and other renewables). Governor Cuomo's New York Sun Initiative puts increased funding toward expanding solar power in the state. New York is definitely facing the sun and striving to be a leader in PV solar energy development.

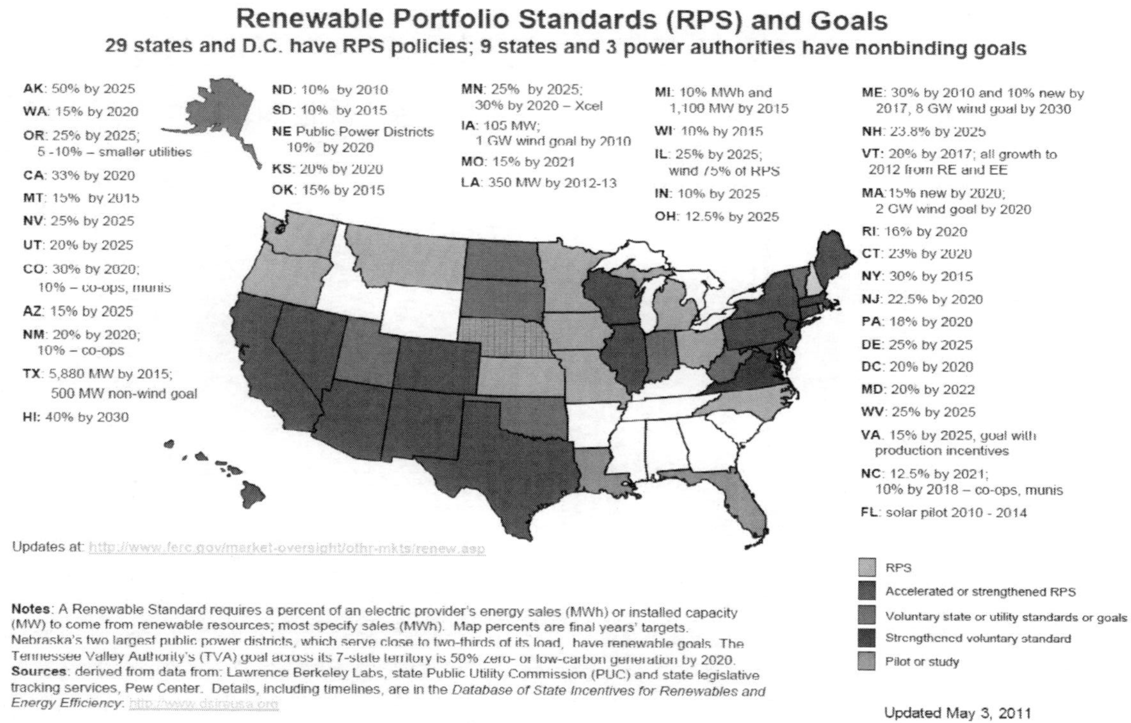

Figure 6. Renewable Portfolio Standards (RPS) and Goals.

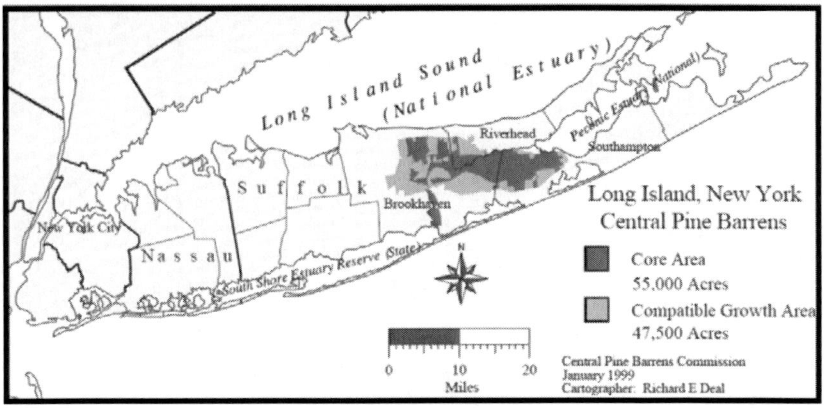

Figure 7. Map of Long Island, New York, Central Pine Barrens.

On Long Island

LIPA, the state-owned utility for all of Long Island, has been a leader for solar development. In 2000, LIPA began a forward-leaning program of incentives for residential and business-scale solar projects with capacities between 50 kilowatt (kW) and 3 MW. LIPA was very bold in issuing an RFP in 2008 for projects that could deliver 50 MW of solar PV. This was the first of its kind in New York state. LIPA has also developed a program for solar feed-in tariffs, to standardize integration to the grid. This is another first-of-its kind program in New York (Deering 2012). LIPA is also consistently ranked in the top 10 utilities in the country in total PV solar capacity (SEPA 2012).

Local Environmental Issues

There is a very strong awareness of environmental issues on Long Island. The very fact that it is an island amplifies the notion that its environmental resources are well-defined. The natural environment is not boundless, as it may seem on the mainland. Instead, its environmental resources are very finite as they extend into the sea in all directions. In addition to this, Long Island rests atop a sole-source aquifer, which means that nearly all of the drinking water on Long Island comes from wells in the ground. The people of Long Island, in general, have a heightened concern about pollution of their waterways and protection of their limited natural resources.

Evidence of this is that the combined funding levels—from town, county, and state laws—for preservation and conservation of natural resources in Suffolk County (the eastern two-thirds of Long Island) is $2.2 billion since 1977, which is more than that of 45 other states (Amper 2012).

Another seminal environmental achievement for Long Island was the legal protection of the Central Pine Barrens, the island's largest natural area and last remaining wilderness. The New York State Long Island Pine Barrens Protection Act of 1993 identified core preservation areas and established the Pine Barrens Commission with jurisdictional authority to manage compatible growth areas (PBC, 2012). The law has been very successful and serves an important feature of Long Island's acute environmental sensitivity.

Summary

At first glance, it may seem peculiar that one of the largest PV solar arrays in the United States—and in the world—is located on Long Island. However, in descending from a worldwide view to a local perspective, it is clear that the concept of a utility-scale PV solar array on Long Island was not an accidental or haphazard coincidence. Rather, there was a nexus of key factors that were essential to its creation: federal agencies looking for innovative development opportunities, states pushing to incentivize solar energy, and communities wanting clean sources of electric power. These conditions created fertile ground for a project like this to take root.

Project Concept

How It All Got Started

For nearly a decade, LIPA had established a suite of PV solar initiatives geared toward small- scale projects, primarily at the residential level. In wanting to move solar to larger-sized commercial projects, in 2008 LIPA issued an RFP for 50 MW of PV solar power from anyone who could supply it. This was a very bold request, and the first of its kind in New York state.

According to Michael Deering, LIPA's vice president for environmental affairs, LIPA was agnostic about whether that 50 MW could be generated by one single array or if it could be done with several smaller arrays distributed around the island. The solar RFP was an open question, and they waited to see what the industry's responses would be. Nevertheless, given the resource constraints on Long Island and the large volume of power they were requesting, LIPA generally expected that they would get those 50 MW in a distributed manner. The RFP opened in April 2008 and closed that following August. LIPA received a total of 35 proposals. Unsurprisingly, most of them were for distributed development of the 50 MW.

Seizing the Initiative

Naturally, one of the biggest resource constraints for PV solar is the availability of land. Generating a lot of PV solar power requires a lot of acreage. On a densely populated island with very finite space, there were few tracts of land that could host a very large array in a cost- effective way. Recognizing this, LIPA released another RFP to identify any land owners on Long Island who would be interested in hosting a PV array. LIPA made the list of potential land owners accessible to the respondents of the solar RFP, but not vice versa (Deering 2012). Again, the majority of the land owners could only support small portions of the 50 MW projects.

Soon after the solar RFP was publicized, both LIPA and BP Solar recognized that one of the single largest land-owners was Brookhaven National Laboratory. In separate calls that happened to be around the same time in May 2008, both LIPA and BP Solar contacted the lab. LIPA made a call to Dr. Looney, of BSA, and BP Solar called Bob Gordon, director of the Business Management Division for DOE at Brookhaven National Laboratory. In both cases, they posed the question, "Would Brookhaven National Laboratory be willing to host a solar farm?" (Looney, Deering, Gordon, Chandler 2012).

Using Federal Lands

Discussions began immediately at the lab between BSA and DOE, who analyzed how such a project might fit the mission of the DOE and serve the interests of the lab. In June, LIPA published on its website a formal letter from DOE. The letter proposed the possibility of using federal property at Brookhaven National Laboratory to companies participating in the solar RFP. This was a very attractive prospect for many companies who were looking for sufficient acreage for a large PV solar project.

What was not attractive to many companies, however, were the unusual stipulations that DOE put on its proposal. In order to use the site, a project would have to agree to two key provisions:

- The project must include a research component, in order to fit with DOE's mission for the lab.
- The land-use agreement would be an easement, not a lease or a license.

These conditions repelled nearly all the would-be developers. The research component of the deal was unusual, to be sure. This was a venture for private business to develop solar power for commercial use. Most companies quickly rejected the superfluous costs that the research component presented.

DOE's insistence on using an easement was widely regarded as very unattractive to investors.[5] In general, developers of projects like this aggressively seek out licenses or leases for full control rights of the land during the time of the agreement. This gives them rock-solid rights to use the property, which provide a critical foundation of certainty for their business models.

DOE could not support such a wholesale transfer of control that was associated with a lease or license. It was essential that DOE retained access to the land under the array. In a practical sense, DOE would need to retain access rights in order to manage the research aspect of the project. More importantly, though, in a long-term strategic sense, DOE needed to reserve ownership of the land in case of a change in federal priorities for the site (Easement Outgrant 2010). To the vast majority of developers, such an easement represented an intolerable degree of uncertainty for their long-term business models. The only developer that did not shy away from those stipulations was BP Solar.

LIPA's Choice—Selection Criteria

After thoroughly reviewing the details of each of the 35 proposals, LIPA made its selection on which companies would fulfill its request for 50 MW. The review process was very rigorous in quantitative and qualitative analysis. The selection criteria was based on a number of factors, including the quality of the developer, cost-benefit analysis, location of the site(s), likelihood of obtaining the requisite permits, interconnection logistics, etc. (Deering 2012).

One of the key aspects of each proposal was the ability of the project developer to demonstrate site control; that is to say that the developer had (or could obtain) total control over the proposed site. Just like the developers, LIPA wanted to minimize or nullify those risks associated with land use in order to safeguard its power purchase agreement. Because BP Solar was the only company willing to take the business risks associated with DOE's stipulations, it was the only proposal that incorporated land use at Brookhaven National Laboratory (Deering 2012).

It was important that there was only one company that was willing to work with DOE. On the one hand, it was easy for DOE to make a commitment to BP Solar, at least in terms of entering negotiations with them and nobody else. On the other hand, that relationship demonstrated to LIPA the requisite level of site control for the proposal. In other words, if multiple companies had submitted proposals that claimed they could use land on Brookhaven National Laboratory, it would have demonstrated that none of them really had any sort of commitment from DOE. This in turn, would have raised the level of uncertainty for LIPA to an intolerable level, and they would have rejected those proposals for a utility-scale array (Deering 2012).

In consideration of all these factors, LIPA ultimately selected two companies, BP Solar and enXco. BP Solar was awarded a contract to construct a single array that would produce 32 MW.[6] The other 18 MW would come from enXco, who would develop a series of distributed arrays built over large parking lots in several places around Long Island.

Power Purchase Agreement

Upon selection, BP Solar established the "Long Island Solar Farm, LLC" as the legal representation of the solar farm as an entity unto itself. LIPA negotiated the terms of a power purchase agreement (PPA) with LISF, LLC. to provide the energy produced and renewable energy credits (REC) from the solar farm. The costs to LIPA under the PPA for the energy produced by LISF, LLC. total approximately $298 million, which include the costs of interconnection. The time horizon of the PPA is 20 years from the date of in-service, which was Nov. 3, 2011.

The results of the PPA amount to about $0.60 per month for the typical residential customer. According to LIPA's Chief Operating Officer, Michael Hervey, "This agreement not only allows us to provide clean energy to our customers, but it also delivers price stability for our customers in an energy market where oil and gas prices remain volatile" (LIPA 2011).

Shared Investment—LISF, LLC

BP Solar could have done this project on its own but chose to bring in a partner as an investor. Once the PPA was signed and the project design was about 30% complete, the project was stable enough to bring an investor on board. MetLife became that partner and served as a passive hands-off investor with no other stake in the project other than its financial investment. Even though MetLife owns the majority stake of the Long Island Solar Farm, they are still an insurance company. They do not have the engineering, procurement, or construction wherewithal to take an active role in the project. BP Solar, therefore, serves as the leader and primary resource of consideration, according to the terms of the easement (Chandler 2012). In order to accommodate the dual ownership, the solar farm itself is established as a limited liability company, officially entitled the "Long Island Solar Farm, LLC."

PROCESS

No Easy Easement

The first and most important aspect of negotiations for this project was the development of the easement agreement between BP Solar and DOE. As stated before, the easement was the best of three options for DOE, and the worst of the three for BP Solar. Lou Sadler, DOE's legal counsel, explained the basic pros and cons of each from the DOE's perspective.

Lease—The basic terms of the lease included things intolerable to DOE:

- DOE would lose control of the site for the time period of the agreement
- Any consideration would have gone straight to the U.S. Treasury

* Major Milestones

	Date	Action	By	For
2008	April 22, 2008	LIPA Issues a Solar RFP	LIPA	Anybody
	May, 2008	BP Solar approaches BNL with an interest to use BNL land; LIPA also approaches DOE with interest for land use at Brookhaven	BP Solar & LIPA	BNL & DOE
	June 30, 2008	DOE/BHSO Submits Hosting Letter - Proposing Access to BNL	DOE/BHSO	RFP Applicants
	August, 2008	Proposals Due to LIPA	Industry	LIPA
	November, 2008	LIPA Selects BP Solar's Proposal - PPA Signed	LIPA	BP Solar
2009	April, 2009	Presentation to Community Advisory Council	DOE - BHSO	Community
	June, 2009	Real Estate Appraisal Published	MacCrate Associates LLC	BP Solar
	September, 2009	Solar Easement Draft Published	DOE	BP Solar
		Appraisal Value Negotiation Meeting	BHSO, BNL & BP Solar	
	December, 2009	NEPA: Environmental Assessment Complete	DOE	BP Solar
		FONSI	DOE	LISF
2010	January 20, 2010	Easement Granted	DOE	LISF
	February 3, 2010	MOU Signed	DOE & BP Solar	LISF
	October, 2010	Easement Management Plan Signed	DOE / BHSO	LISF
		Construction Activities Begin	LISF	DOE/BHSO
	December, 2010	Begin Development of CRADA 1 (Cooperative Research and Development Agreement) for Instrumentation of the LISF	BNL & BP Solar	LISF
2011	August, 2011	Liability Insurance Issued		
		Building Code Compliance Approval	Sigma Energy Engineering	BP Solar
	September, 2011	Begin Developmnet of CRADA 2 (Cooperative Research & Development Agreement) for the Dedicated Research Array	BNL & BP Solar	LISF
		NEPA Review of Research Array Site Complete	DOE	BP Solar
	November 3, 2011	Flip the Switch: Full Commercial Operations Begin at the LISF	BP Solar	LIPA
	December, 2011	Easement Management Plan Signed	DOE / BHSO	BP Solar
		CRADA 1 and CRADA 2 are executed	BNL & DOE	LISF
		Press Release: BP Solar Will Go Out of Business	BP Solar	BP
2012	Spring 2012	Continue to Develop CRADAs 1 and 2, Transition Ownership of the LISF	BP Solar	BNL
	November, 2012	Begin Construction of Research Array	BP Solar	BNL
2013	March, 2013	Flip the Switch: Research Array is Operational	BP Solar	BNL
	November 3, 2031	Easement Expires	DOE/BHSO	LISF

Figure 8. Generalized Timeline of Development for the Long Island Solar Farm.

- No opportunity for scientific research
- Would greatly extend the timeline (involving formal public comment periods).

License—The basic terms of a license were similar in many ways to those of a lease:

- Both leases and/or licenses were attractive to investors because they are nice, neat, and locked in for long periods of time. This reduces risk in the analysis of the financiers.

Easement—This was the only way DOE was willing to move forward with the project:

- Easements are not typically used for projects like this (leases or licenses are predominantly used by investors).
- Easements generally allow for someone to have very limited use of a parcel of land; a common example is for the use of telephone poles and/or electric power lines to cross a particular area.
- The rationale was that, at its core, there wasn't much difference between the LISF project and the common application of an easement for development of transmission lines; the solar power array would be constructed over land that was otherwise unused.
- DOE also insisted on the use of an easement because they would retain:
    - Full ownership and access to the land
    - The compatibility of the project with DOE's mission, through the availability of in- kind consideration to be applied directly to the research agenda
    - The ability to meet the condensed timeline of the project, as set by LIPA.

The easement negotiations were somewhat tricky. BP Solar very much wanted the agreement to sound as much like a lease as possible, of course, to minimize their risk. In contrast, DOE insisted that it must retain access to and control of the land, and could not risk relinquishing it in a lease or license.

Consequently, as the two sides negotiated to common ground, they tried to narrow the gap of risk between a lease and an easement. Richard Chandler, project manager for BP Solar, described the negotiations:

> Ultimately, it's not the label at the top of the document, but the rights and obligations outlined in the document that are most critical. So long as we were okay with the responsibilities and tenets of the agreement, we were willing to move forward. The feeling, of course, was mutual on the other side of the table. But, we wanted to narrow the language of the agreement to ensure, as much as possible, that if BP Solar dedicates its resources, then the land would not be taken away (Chandler 2012).

Mr. Chandler alludes to the key feature of the negotiations—trust. In other words, in order to make a final deal, that gap of risk between a lease and an easement needed to be spanned with mutual trust that couldn't be captured per se on a legal document.

On one side of that gap, BP Solar needed to trust that DOE would not pull the land out from under the array for any unexpected reason in the next 20 years at a minimum (the term of BP Solar's PPA with LIPA). In practicality, the solar farm site represents a small portion of all available (or unused) lands at the lab. Thus, it would be extremely unlikely that DOE would have a reason to have to go through the massive undertaking of dismantling the solar farm and either repurposing that specific land for other development or restoring it to its natural state. Such a scenario was so unlikely that BP Solar was willing to trust DOE's assurances that they'd let the Solar Farm exist on that site for the duration of the PPA. This is a strong measure of good faith, as the easement explicitly authorizes DOE to do what it wants with the land beneath the solar site at any time (Easement Outgrant 2010).

On the other side of the gap, DOE had to trust that BP Solar would uphold its end of the deal and provide in-kind consideration to develop the dedicated research array. BP Solar wanted—and was afforded—specific language in the agreement that enabled them to liquidate the value of consideration and write a check to the U.S. Treasury (Gordon, Lofaro 2012 and Easement Outgrant 2010). This was not in the best interest of DOE or the lab. If BP Solar did that, it would completely undermine the research component of the solar farm, and disassociate the entire project from the mission of DOE and the lab. Nevertheless, and despite the significant risk, DOE trusted that BP Solar would follow through with its commitment to provide consideration instead of cash.

In the end, the keystone of the whole easement—and consequently the entire solar farm project—is that level of trust over crucial interests at the core of the agreement.

Reinforcing Commitments, Insuring Trust

As a compromise, written explicitly in the terms of the easement, BP Solar insisted on retaining the ability to write a check to the U.S. Treasury for the balance between the estimated total value of consideration and the actual in-kind consideration they could deliver. To illustrate, if BP Solar was committed to a package of in-kind consideration (of tools, equipment, special expertise, etc.) valued at $100,000 but only contributed $70,000 worth of necessary resources, then BP Solar would be able to write a check to the U.S. Treasury for the remaining balance of $30,000.

This posed a significant liability for DOE and the lab because, at any time, BP Solar could choose to liquidate its commitment of consideration and pay it to the U.S. Treasury, which does not directly benefit the interests of DOE and BNL (Lofaro 2012). It would undermine the research component of the project, which was the hallmark feature and the essential viability of the project from DOE's perspective.

In order to ensure that BP Solar followed through, DOE insisted on an up-front commitment of funds from BP Solar that would be dedicated expressly to the development of the dedicated research array. This was a commitment that took some leveraging of BP Solar, who didn't think it was necessary for DOE to insist on such a down payment. Nevertheless, DOE wanted a tangible commitment to protect the development of the research array if BP Solar was inclined to back out of the project for some unforeseen reason.

Despite their verbal assurances that they would follow through in providing consideration, BP Solar ended up having to back out of their long-term commitment. The company declared it was going out of business soon after the LISF was built but before the design began on the dedicated research array. Were it not for this down payment to BSA, BP Solar would have been able to write a check to the U.S. Treasury for the difference. This would have technically fulfilled their legal obligations in the easement and enabled BP Solar to walk away (Sadler 2012). Of course, that disastrous scenario would have left DOE and BNL without a dedicated array for solar energy research.

BSA currently has that down payment of $2.3 million in holding (Gordon 2012). That investment is, at the time of this publication, driving BP Solar to fulfill its commitment to developing the research array before they completely decouple themselves from this project.

PROJECT DESIGN

The Use of Consideration

As stated previously, in order for a commercial project to be built on federal property at a research institution, it must incorporate a research component that would effectively support DOE's mission for the lab. Richard Chandler, project manager for BP Solar, recognized this early on. He saw it as an opportunity that could benefit not just the lab but also his own company. He also recognized that the federal agencies weren't as interested in monetary compensation as the landowners typically were in other solar projects. Under the terms of the easement, BP Solar worked very closely with BSA and DOE to develop the definitions of "consideration," which would serve as the exchange of in-kind contributions to a Cooperative Research and Development Agreement (CRADA). Consideration refers to compensation given by one or more parties to an agreement that can be in the form of goods, services, and/or money.

According to the easement, BP Solar is obligated to deliver consideration to DOE in return for the easement to use the land. The total valuation of the consideration in dollar terms is approximately $6 million over the 20-year easement. This has been estimated and generally agreed upon as the aggregate value of all equipment, labor, construction expenses, maintenance issues, land use, and costs associated with the research agenda (Gordon 2012). So, instead of BP Solar paying a full rent check to DOE for use of the land, BP Solar would provide other resources in the way of materials and expertise. Any balance left over would be paid by BP Solar directly to the U.S. Treasury. DOE, of course, prefers that the consideration be applied on site at the lab (Easement MOU 2010).

As Richard Chandler describes, "The use of consideration was a way to find common ground and create a mutually beneficial exchange of goods, services, and resources. This was a natural fit for both BP Solar and Brookhaven. Both BNL and BP Solar were very interested in Solar (research and development). The bottom line is that developers need to approach a project like this, dealing with multiple entities, particularly with federal agencies, with an appreciation for what the other side is looking for" (Chandler 2012).

Two-Pronged Approach

The quintessential research component of the Long Island Solar Farm was formally codified in the CRADA. This agreement was broken into two parts, CRADA 1 and CRADA 2, each with a different research focus:[7]

- CRADA 1: The inclusion of data collection instrumentation on the LISF
- CRADA 2: Development of a dedicated research array (NSERC)

TECHNICAL DESIGN

Overall Description

The system is a large-scale commercial solar photovoltaic array of approximately 37 MW (direct current), and covers approximately 200 acres of the BNL federal site. Electricity generated by the system is connected into the regional utility power grid. The system is comprised of individual solar modules. Approximately 164,000 modules are used, and the current design has modules arranged into sub-arrays consisting of four vertical rows of six modules. The 24 modules are attached to a steel I-beam support structure that is anchored with two steel I-beam driven posts. These posts are driven approximately 15 feet (ft.) [4.6 meters (m)] below the surface of the soil. Strings of modules are wired together in series. Fused combiner boxes with disconnect switches connect strings of modules and lead to 32 concrete equipment pads of about 100 square feet each, placed around the array. Each equipment pad contains two inverters and one megavolt ampere (MVA) transformer. The wiring between the modules and equipment pads are trenched. Conductors combine at 15 kV outdoor, metal-clad, switchgear on a 250 square-foot concrete pad. A 13.8/69 kV transformer, with associated circuit breakers disconnect switches and a small (12"x24") control enclosure mounted on pads totaling 2,000 square feet, then steps up the power and supplies it via cable to the Long Island Power Authority 8ER Substation, located on the other side of the Long Island Rail Road (LIRR) track from BNL. The 69 kV lines, including three conduit lines, [one for fiber optics and two for power distribution (one active and one spare)], run under the Long Island Railroad tracks (Easement Outgrant, 2010).

Technology

The array is polycrystalline photovoltaic cells on a fixed array, oriented due south and tilted at 27 degrees. This technology and configuration optimizes the array for maximum production output at its particular geographic location and also during the summer months, when loads on Long Island are peaking. This was also the most efficient technical design in terms of cost- benefit analysis (Chandler 2012).

Patchwork Design

The easement authorized BP Solar to use oddly shaped portions of the property. At first glance, this arrangement appears to be very random. It was, however, the result of very deliberate environmental considerations and impact mitigations.[8]

Capacity
32 MW (ac) delivered to LIPA

PV Module Specifications
BP Solar Models
BP3225T and BP3230T
Number: 164,000
~23 modules/string,
~7100 strings
~290/inverter

Technology:
Polycrystalline Silicon

PV Module Installation:
Ground mounted,
24 panels/rack,
~7000 racks
Module Tilt: 27°Fixed;
Array Azimuth: 180°

Inverter Power Blocks:
Manufacturer: SMA Sunny Central 630HE

Net Nameplate Rating:
1.25 MW

Figure 9. System Specifications.

This patchwork of six separate areas presented significant challenges for design. It required BP Solar to spend a lot of time reconfiguring the technical blueprint of the array, which increased the overall design cost. By comparison, the large-scale arrays built in the open deserts out West are typically laid out in nice rectangular blocks with lots of room to spare. On Long Island, however, it was a constant challenge to pack in as much power onto parcels of land that were both incongruous and very irregularly shaped (Chandler 2012).

Transmission

It was very fortunate that LIPA already had an existing substation immediately adjacent to Brookhaven National Laboratory. As shown in Figure 10, even more convenient was that the substation was adjacent to the southeast corner of BNL property, which was otherwise unused. This made the southeast corner of BNL an ideal site for the array. These serendipitous conditions completely eliminated the need to build attendant transmission infrastructure for the project. For all intents and purposes, they could literally plug the whole thing right into the grid.

Though the distance for transmission was negligible in terms of its impact on the whole project, it was not zero. Approximately 900 ft. (274 m) of transmission cable were installed between the BNL property line and the LIPA substation 8ER point of connection to the grid. The cable, consisting of two conduit lines (one active and one spare), had to cross the LIRR tracks between the lab property and the substation. The stretch of transmission line that crossed the LIRR easement was constructed belowground using horizontal directional drilling (EA-1663 2009).

When asked whether BP Solar would have considered building such a large array if that substation was not there, Richard Chandler explained, "It would be difficult to answer. Undoubtedly, having to create transmission infrastructure would have added another layer of complexity to the project, but in the end, the answer to this question would come down to a standard cost-benefit analysis." The phrase, "another layer of complexity" may be an understatement, depending on how far a transmission connection would have to be. Given the location of the lab property, surrounded by state-protected wilderness, endangered species habitat, and very expensive private lands, it would have been another major project in its own right to design, site, permit, and build attendant transmission infrastructure. It certainly would have added substantial costs of time and money to the overall project (Chandler 2012).

Figure 10. Satellite View of the LISF and NSERC at Brookhaven National Laboratory.

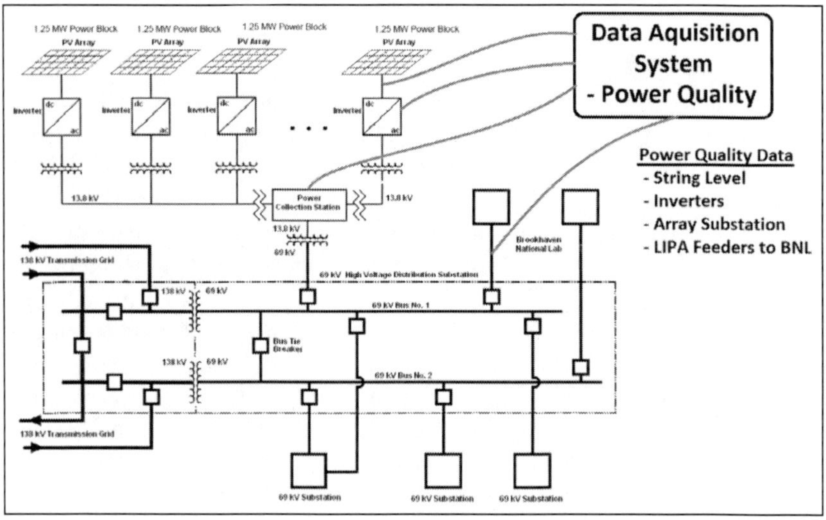

Figure 11. Grid Integration Diagram: Long Island Solar Farm.

Interconnection to the Grid

The LIPA substation required only very minor modifications to connect the LISF to the energy distribution grid. The substation modifications included the addition of two 69 kV disconnect switches, a 69 kV gas circuit breaker, 69 kV potential transformer, revenue metering, metering potential transformers and current transformers, and related control and protection relaying (Figure 11). All modifications occurred entirely within the substation footprint (EA-1663 2009). There were no significant technical challenges associated with the interconnection with the grid. Once the design was 30% complete, LIPA, BP Solar, and New York Independent System Operator (NYISO) conducted an 18-month study of the solar farm's impact on the grid and found there to be no significant issues. This was the first solar project integrated into the NYISO, using their standard forms and procedures for integration (Chandler 2012).

Vulnerabilities

Typically, solar power plants are most vulnerable to theft and vandalism. This is, in part, because they're typically in remote locations and require little-to-no operational supervision. Most stations are protected by fencing of some sort and many are monitored remotely with cameras (Chandler 2012).

The location of the Long Island Solar Farm insulates it from this type of vulnerability. The LISF is very difficult to access, in part because it is surrounded by the Pine Barren wilderness and also because it is located on the lab property, which is fenced-in, guarded, and patrolled.

As a consequence of its location, the Long Island Solar Farm is vulnerable to more extreme weather conditions than large arrays in other places. Long Island is prone to hail-producing thunderstorms in the summer, hurricane-force winds in the fall, and heavy snows in the winter.

In June 2012, for example, a portion of the array suffered a lightning strike during a particularly intense thunderstorm. Though the LISF has a 50-ft. lightning rod in the center of the array, the lightning hit one of the racks, busted dozens of fuses, and fried some electronics. This vulnerability to natural conditions represents a potential risk of high operational and maintenance costs. It also serves, however, as a uniquely valuable opportunity to research and model the effects of such varying conditions on a large array.

Public Engagement

Proactive from *(before)* the Start

BP Solar put a lot of effort into making a good first impression with the community. Soon after LIPA selected their proposal, BP Solar organized an open house at a community center near the lab. The open house included all the proponents (LIPA, DOE, BP Solar, and BSA) and was open to everyone. The open house created a very good first impression, and the project did not receive any initial resistance.

Subsequent meetings and updates weren't standardized, but they were frequent enough to sustain the engagement between affected groups: DOE, BP Solar, LIPA, and the local community, elected officials, and environmental activist groups. BP Solar was very engaged from the onset. The company:

- Gave presentations to BNL Executive Committee and the Brookhaven Community Advisory Committee
- Visited all the local chambers of commerce
- Participated in BNL Summer Sundays (a series of open houses at the lab where the community is invited onto the lab for tours and educational programs); BP Solar's participation in this was not specifically about the solar farm project but more of a "PV 101" about solar energy to increase awareness of the technology
- Dispelled rumors and misinformation by posting billboards near/on site to explain to passersby details about the project
- Distributed flyers to everyone who was on or near the transportation route associated with the construction of the solar farm
- Solicited ideas from the public on how to improve the project and incorporated many of their suggestions into the project's development (Chandler 2012).

More broadly, although there was generally not a lot of media attention given to the project until it was near completion, a few reporters in the regional print media consistently published updates on the development of the solar farm. This was helpful to keep the public aware of the progress and status of the project. In general, the public engagement strategy was generally to lean forward and get out ahead of any issues. This active approach to engaging with the community was instrumental in defining the public message, communicating progress, and minimizing the grounds for backlash from political or special interest groups (Gordon 2012).

Richard Chandler concurs with his description of BP Solar's outlook on communication. He explains that the biggest takeaway regarding public relations was that, "It is really easy for misinformation to get out in front of a project, which could have negative and/or crippling effects on its development. We got out there early and often to get ahead of any of that to set the tone and control the message; to make sure the public was getting accurate, forthright information about the project's development." Both DOE and BP Solar recognized that this project could not go on without public support. Their proactive strategy of engagement ensured the public was a part of the development process, which significantly enhanced its chances of success.

SITING AND ENVIRONMENTAL ISSUES

Environmental History and Current Conditions

Superfund Site

All of BNL is classified as a Superfund site and is listed on the National Priorities List, which is administered by the Environmental Protection Agency. This classification makes BNL eligible for funding relating to environmental cleanup and mitigation of the effects of hazardous materials. This is primarily because of the types of pollutants that were discarded and the unique aspects of some of the decommissioned laboratory facilities on site. Brookhaven National Laboratory's Superfund classification has no direct impact on the easement, land use, or the LISF project. The only exception is that some parts of the BNL property are classified as off limits or unusable (Gordon 2012).

Key Environmental Interest Groups—A Snapshot

Citizens Campaign for the Environment (CCFE)

Adrienne Esposito, Founder and Executive Director CCFE:

- Is a lobbying entity
- Has no legal authority
- Created a groundswell movement to support the project
- Is an advocate of LIPA's perspective
- Thought if this project did not happen now, then it would never happen—or a polluting alternative would eventually be built instead.

The Long Island Central Pine Barrens Society

(aka "The Pine Barrens Society" or "LIPBS")
Richard Amper, Executive Director
LIPBS:

- Is a lobbying entity
- Has no legal authority
- Felt the environmental benefits of this project were not worth the environmental costs of destroying Pine Barren lands
- Never endorses development but only actively opposes a development plan if it threatens to negatively affect water quality and Pine Barren ecosystems
- Has three basic levels of opposition:
    - Level 1—the society publishes a statement in opposition to a project.
    - Level 2—the society puts pressure on appropriate authorities to oppose a particular project.
    - Level 3 (litigation)—the society will file suit against a developer for violating the law. Projects very rarely come to litigation (approximately two cases per year) (Amper 2012).

DOE was not legally obligated to work with the LIPBS.

Central Pine Barrens Joint Planning and Policy Commission
(aka "The Pine Barrens Commission" or "PBC")

- Is a state entity with governing authority over the Pine Barren lands, according to the New York State Long Island Pine Barrens Protection Act of 1993
- Does not have jurisdiction over federal land on BNL property, though the Pine Barren habitat extends onto it.

Steven Jones was commission chairman at the time of development and strongly opposed the project as counter to the DOE's mission (Amper 2012).

Long Island Native Plant Initiative (LINPI)

- This group generally resisted the development of the LISF.

They insisted on protecting the site from invasive species of plants and that the undergrowth of the LISF would be planted with native grasses only.

Divided Resistance

The environmental community was at odds over this project. On one hand, some people viewed the solar farm as a clean source of renewable energy that was better than an alternative source of power that would likely generate more pollution. This broad perspective of the project had its loudest voice in the Citizen's Campaign for the Environment (CCFE).

On the other hand, there were environmental activists who viewed this project as being destructive to the environment on which it would be built. This perspective's chief spokesman was Richard Amper, executive director of the Pine Barrens Society. He vehemently disagreed that renewable energy projects should come at such a high cost to the environment. He states, "The Society has worked with CCFE for twenty years and has agreed on just about everything, but this project posed the first dramatic rift between our organizations. Though we were always mutually very respectful, Adrienne [Esposito] and I had some challenging phone calls. We ended up agreeing to disagree" (Amper 2012).

Fundamental Environmental Debate

> *"The Long Island Solar Farm destroys some aspects of the environment in order to preserve others."*
>
> ~ Richard Amper

The LISF indeed represents a cost of more than 160 acres of forest. Considering the cost, however, the carbon offset of the solar farm is at a much higher rate than the amount of carbon that area of forest could sequester. Plus, 100% of the land used in the project received some type of mitigation and an equal amount of lands were protected elsewhere, both on and off the property (Green 2012).

Richard Amper's perspective was that this project set up a false choice between "preserving natural ecosystems or building a solar energy farm." He states:

> This project extracted an environmental cost to build an "environmentally friendly" energy resource. The damage caused to the environment by the construction of the solar array is not offset by the solar farm's environmental benefits. The notion that we can develop renewable energy, but it has to be at a cost to the environment is a false premise, and that's what this project represents. LIPBS supports renewable energy and solar power, but not at the expense of Pine Barren ecological systems.

The alternative to the single-source utility-scale array would have been aggregating the necessary 150 acres of land from other areas, whether they be onsite at BNL or distributed throughout the rest of Long Island. LIPBS, therefore, much prefers the distributed model of solar development like the enXco project that paralleled the LISF development, according to LIPA's 2008 RFP (Amper 2012).

Legal Lines in the Sand

The environmental groups did not have any legal jurisdiction over the site. However, that didn't mean they were powerless to stop the project. Groups like the Pine Barrens Society could very well have killed the project by delaying it. They could have filed suit, requested an order of restraint, or drummed up popular resentment against the solar farm. Any of these contingencies, by themselves or in combination, could have significantly delayed the project. If delayed long enough, the project would have failed.

This project proved to have an interesting dynamic in the eyes of environmentalists. The stark contrast of perceptions for and against the solar farm created a tension in the environmental community that prevented a strong coalition of opposition from forming. This defeated any chance for environmentalists to create the political or social conditions that could suffocate the project.

Nevertheless, DOE was very forthright in dealing with the environmental concerns of the project. BP Solar and LIPA also took a forward-leaning approach to mitigating environmental impacts. In the end, the loudest sources of opposition (i.e., the Pine Barrens Society) were largely satisfied with the mitigation package from DOE, BP Solar, and LIPA.

Engaging the Opposition

Richard Chandler states, "By and large, most environmental groups were supportive of the project as a clean source of renewable power. The Pine Barrens Society, however, was the staunchest in its opposition. Our intent was not to oppose them but rather engage with them. We solicited their input and used their feedback to improve our project and mitigate its environmental impacts."

DOE established a similar disposition in dealing with environmental opposition. Bob Gordon explained that, "Right from the start we knew that the Pine Barrens Society would be the most vocal opponent to the plan. We met with them early and got them involved in the process right away, and made every effort to incorporate their ideas and suggestions."

This strategy was effective in engaging the project proponents with the LIPBS; however, Richard Amper disagreed. "BNL and DOE have always been very forthcoming with their plans. BNL (DOE) is the largest single owner of Pine Barrens habitat on Long Island. [The BNL-LIPBS] relationship had always been generally good up until the LISF project. The project had a cooling effect on that relationship" (Amper 2012).

Mitigation Package: Natural Resources Benefits

DOE, BSA, LIPA, and BP Solar were very interested in honoring the environmental considerations of opposition groups as much as possible. In working with the LIPBS and the Pine Barrens Commission, the project proponents put together a natural resources benefits package to mitigate the environmental impacts of the solar farm. The key tenets of this package include (CCFE, 2012):

- LIPA provides \$2 million for open space preservation within the Central Pine Barrens region
- DOE preserves an additional 51 acres of property that builds upon the 500-plus acres it previously preserved in the 1990s
- BP Solar provides \$75,000 for ecological habitat, research, and restoration.

\$2 Million from LIPA

At first glance, this appears to be an ostensible "mitigation fee" put on LIPA. This is a misperception, according to Mike Deering. He explained that, "The \$2 million was part of the regulatory process, as administered by the New York State Pine Barren Commission. They have jurisdiction over the development of the Pine Barren region, but not direct authority over the federal property. The commission suggested a 1:1 preservation match, and \$2 million facilitated that. LIPA honored their suggestion and appropriated the money out of an existing fund for community benefits. That money was then dedicated for the purchase of lands for preservation."

Richard Amper agreed that was an important piece of mitigation. "Those funds essentially went through Brookhaven town and were subject to public input. The LIPBS contributed guidance on selection of lands for preservation. The final decision—to buy land in the Carmans River Watershed—reflected our input."

51 Acres from DOE

The Pine Barren Society had no legal influence over the project, but DOE worked to get the LIPB on board as much as possible. Richard Amper agrees by saying, "DOE, to their credit, were consistently and sincerely candid, accessible, and professional in their dealings with [the LIPBS]. DOE took us on a tour of the facility and consulted with us for their site selection. DOE clearly wanted to minimize our unhappiness about the project. DOE was very upfront about their plans. Right from the start they set an inclusive tone." He also included a caveat, though. "Though they were inclusive, DOE made it clear very early on that this project was going to go forward, regardless of [LIPBS's] analysis that solar energy should be developed elsewhere and without damaging Pine Barren ecosystems."

DOE made concessions to the society throughout the process. One of these measures was to commit other portions of the BNL property to preservation. Thus, DOE classified 51 acres of wooded property as unavailable for future development by the lab.

It is important to note that this is not a binding agreement; DOE is in no way obligated to abide by that self-imposed measure. It has, nevertheless, incorporated that preservation area into its official long-term development plan.

$75,000 from BP Solar

This lump sum went to Long Island Native Plant Initiative (LINPI), which is a cooperative volunteer not-for-profit effort of more than 30 organizations, governmental agencies, nursery professionals, and citizens. It was founded in 2005 by Polly Weigand, technician for the Suffolk County Soil and Water Conservation District. LINPI continues to pioneer the collection and banking of seed from more than 35 native grasses, herbaceous plants, shrubs, and trees on Long Island. The grant from BP Solar was intended to go toward LINPI's ecological habitat, research, and restoration.

This was, more or less, a gift from BP Solar. According to Dr. Green, "The long-term benefit of this contribution, though, is that BNL will continue to have access to a source of native seeds that can be used in various projects on the Lab property (BNL would, of course, purchase that seed at market price). This will benefit vegetation programs at BNL, including on the land beneath the solar arrays."

Siting the Array in a Sensitive Environment

In addition to the marquis features of the natural resources benefits package, DOE and BP Solar put a lot of effort into making various other environmental considerations that shaped the final footprint of the solar farm.

The map in Figure 12 gives a broad perspective of some of the major environmental constraints on the lab property. The areas that were already open for development (shaded in pink) are very small and scattered and would not be able to efficiently support such a large system. The areas surrounded by the red borders indicate preferred or "core" Pine Barren ecosystems or compatible growth areas (CGA). In addition, the 1,000-ft. buffer established to protect the endangered tiger salamanders creates a random assortment of exclusion zones. This map reveals how the development of the solar farm resulted in a patchwork of six parcels of land (shaded in blue).

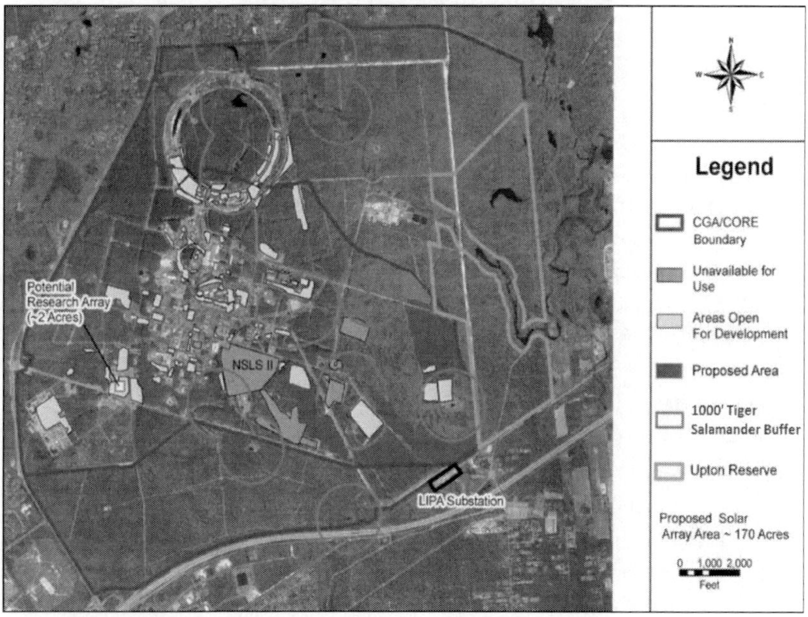

Figure 12. Proposed LIPA solar project.

Figure 12 does not show all of the considerations that went into the site selection for the solar farm. It does, however, give a sense of the major constraints and the limits for development. The full suite of criteria demonstrates the complexity of the project and the high degree of concern that DOE, BSA, and BP Solar had in minimizing then environmental impacts where possible.

The following is a list of considerations made during site selection and design of the project:

- Considerations for selecting location:
    - Avoided core preservation area of Pine Barrens
    - Proximity to LIPA substation
    - Avoids or reduces environmental and cultural resource impacts
    - Utilizes already cleared or previously disturbed areas
    - Limits impacts on BNL operations (away from utilities, traffic, and future science projects).
- Land type (total area is approximately 200 acres):
    - 35 acres of existing already-cleared lands
    - 5 acres of former tree nursery
    - 98 acres former World War I areas, with various levels of restoration
    - 62 acres that were already disturbed prior to World War I
- A portion of project area was moved to avoid 14 acres of higher-quality Pine Barrens habitat in CGA of the Pine Barrens
- The irregular layout of six separate parcels of land minimizes environmental impacts
- The project completely avoids development within core preservation area (CPA)
- Avoids wetlands and maximizes buffers around tiger salamander habitat
- Improves a small pond that is a breeding area for tiger salamanders
- Native grasses were planted to restore cleared lands
- Removes invasive plants and BNL actively prevents the establishment and spread of invasive species into the CPA
- No significant impact on groundwater
    - Total annual water use for maintenance less than 500,000 gallons
    - Native vegetation below the arrays will filter precipitation as it infiltrates ground.
- The project is not expected to impact surface water; current flow patterns will stay unchanged
    - Impervious surfaces increase by 10,890 square-feet (ft^2)

- Creates a deer-free area, enhancing habitat for other wildlife, and creating an ecological research opportunity
- Fencing is wildlife friendly, including gaps (approximately 5"x12") deliberately placed at the ground level, every 75 feet along the perimeter; this allows small animals and mammals unencumbered access in and out of the site
- Construction activities were timed to reduce disturbance to birds and wildlife
- Array connections to step-up transformers follow existing roadways and avoid wetlands.

PERMITTING

At first glance, the location of this project on federal property might appear to be a disadvantage. The project would be subject to multiple layers of review at the local, state, and federal levels, which could jeopardize the timeliness of the project's development. In the case of permitting, though, the unusual conditions of the property turned what could have been a disadvantage into a fortuitous benefit for the project.

Land Development at Brookhaven—A Storied History

Prior to becoming Brookhaven National Laboratory, the federal property was originally a military installation called Camp Upton, which opened in 1917. Shortly after World War II ended, Camp Upton was decommissioned, and the property was turned into a research institution under the jurisdiction of the Atomic Energy Commission, the precursor to the U.S. Department of Energy. From 1947 to 1989, the lab used and disposed of hazardous materials at the lab. The Environmental Protection Agency added the Brookhaven National Laboratory site to the Superfund National Priorities List on Nov. 21, 1989, because hazardous chemicals were found in the soil and ground water (EPA 2012).

This high-profile history of the lab site required several detailed reviews of the property through many different lenses. From a military perspective, the land has been closely surveyed for safety purposes, namely unexploded ordinance or hazardous material still buried in the ground. From an environmental perspective, the property has undergone a battery of very detailed reviews of its geological, hydrological, ecological, and biological characteristics. Even from a cultural perspective, the fact that it used to be a military installation has given cause for the State Historical Preservation Office (SHPO) to hold a record of surveys that catalogs military artifacts and important historical sites.

BNL/DOE Land Preservation History

- 2,943 acres preserved (56% of current footprint)
- 2,339 acres donated to NY State Parks – 1973
- 74 acres miscellaneous land transfers
- 530 acres dedicated as Upton Ecological & Research Reserve (10% of land area) – 2000

BNL Land Development over 60 year history

- 1,412 acres (27%) of current 5,265 acres
- Original WW II area – 500 acres (10%)
- Since 1947 – 912 acress (17%)
- Long Island Solar Farm – 200 acres (<4%)
- 30.6 % developed by 2011

All these surveys and reviews over the years created a very thorough library of information with much of the critical information required by the permits. This turned out to be a great advantage for the development of the Long Island Solar Farm because all the pertinent information was readily available and comparatively little field work was necessary. According to Dr. Timothy Green, Environmental Protection Officer for BNL, "We studied the impacts and potential impacts of the project to the smallest degree. It was fortunate for the environmental review that there was already a large amount of detailed information about the site. This is largely because the lab used to be a military base, it is a Superfund site, and its environmental resources have been studied very carefully over the last 80 years."

Federal Permits

A full review was required under the National Environmental Protection Act (NEPA). This project required an environmental assessment, which is the second-most thorough level of review after an environmental impact statement (EIS). DOE was ultimately responsible for the NEPA certification. The formal environmental assessment was completed in December 2009. An official finding of no significant impact (FONSI) was published by the DOE shortly thereafter.

New York State Permits

- **State Environmental Quality Review Act (SEQRA)**—Because LIPA is a state-owned entity, this project was subject to the State Environmental Quality Review Act and needed a SEQRA certification. BNL conducted its own parallel review according to SEQRA, which required coordination between the lab and the utility. Both came to consensus on what was being analyzed and their respective results (Green 2012).
- **New York Department of Environmental Conservation (NYDEC)**—BP Solar worked very closely with the NYDEC regarding appropriate permits to build in proximity to particular species (in this case, key species include the box turtle and tiger salamander).
- **State Historical Preservation Office (SHPO)**—The historical nature of the site, particularly its military heritage, required a permit from the State Historical Preservation Office. The SHPO had to certify it had conducted an archaeological survey of the site.
- **Wild Scenic and Recreational Rivers Act**—This was a combined permit for freshwater wetlands and scenic or "viewshed" impacts. First, BP Solar agreed to do minor restoration of a freshwater pond in the center of the site. The pond was a breeding area for the tiger salamander, which is designated by New York State as endangered (NYSDEC 2012). Working in this area required a freshwater wetlands permit. Secondly, Brookhaven National Lab is very close to the Peconic River, the longest river on Long Island. Because of the potential effect of the solar farm on the wetlands and the viewshed associated with the river, this project needed a scenic rivers permit (Green 2012).
- **Radiological Contamination Site Interagency Agreement**—This was a multi-agency process to get NYDEC, EPA, DOE, and the Suffolk County Government to come to agreement on the status of environmental conditions and their effects on the development and operation of the LISF. Examples of some of the provisions include:
  - The standards to which the site had to be cleaned
  - Prohibitions on removing soils from the site
  - Protections for workers, etc.

- Though it was a challenge to coordinate among local, state, and federal agencies, the process was remarkably efficient. This was, in part, because most of the detailed information about the site was available at the onset, and the group could get right to work sorting out the details of how to manage those conditions. DOE took the lead to shepherd this group through discussions, and the committee reached agreement in less than nine months (Green 2012).

Local Permits

- Suffolk County requires a heavy equipment use and construction permit for projects like this within the county. Legally, though, BP Solar was not obligated to have that permit from the county government as the project was on federal land. Nevertheless, BP Solar obtained that permit from the county as a tangible way to demonstrate in good faith their responsibility as a "good neighbor" (Chandler 2012).

Research Opportunities

Lynchpin for Development

The willingness of DOE to make part of its federal property available to a commercial energy generation facility is predicated on the research component of the project. As described earlier in the concept of design, there are two major areas of research associated with the Long Island Solar Farm, each of which is outlined in CRADAs.

CRADA 1—The inclusion of data collection instrumentation on the Long Island Solar Farm

Key Research Areas:

- Variability and nondispatchability
- Grid integration of PV solar generation
- Environmental and ecological impacts of large-scale PV solar arrays

CRADA 2—Development of the NSERC

CRADA 1: Data Collection at the Long Island Solar Farm

Overview

The general idea for the CRADA 1 research was to develop and maintain a stream of data from the solar array with a much higher level of granularity than what is tracked throughout the industry. Typically, solar arrays record conditions at intervals of approximately 15 minutes. In contrast, this project was designed to collect data down to the minute or even seconds in some cases. The intent is to compile very detailed data to inform statistical modeling of cloud formations, weather patterns, and general environmental conditions. This will enable researchers to analyze and predict conditional effects on the power performance of the array (Lofaro 2012).

Armed with such vast amounts of data, BNL researchers will be on the cutting edge of climate forecasting research, which could result in new ways to accurately predict power capabilities in the next 30 minutes, the next several hours, or even the next day. This could have tremendously positive impacts: first, it would greatly enhance the solar farm's integration to the local grid, and second, it will help pave the way for further development of solar energy in the East by adding more predictability and certainty to investors.

The Problem of the Day

One of the primary problems with solar energy (and other forms of renewable energy in general) is that sunshine is highly variable and difficult to predict; it just isn't sunny all the time. Obviously at night, the solar farm does not produce any electricity, which is very predictable. The real challenge, though, is the unpredictability of the solar resource at any given time on any given day. PV panels operate very simply—when sunlight hits them, they generate electricity. However, when the sunlight is diffused by clouds, snow, dust, or any other condition, then the output of the array fluctuates precipitously.

Consider a partly cloudy day, for example. As clouds pass over the array—or even just a portion of the array—the power output of the system falls off. As soon as the clouds blow over, the array will ramp back up to full power output instantaneously.

These very steep ramp rates pose a significant challenge to utilities, who cannot predict the output of the solar farm. This makes it impossible for utilities or system operators to rely on the power coming off the array for any meaningful period of time. Figure 13 demonstrates the amount of sunlight received by the LISF in a predictable bell curve over the course of a sunny day. This is a stark contrast Figure 14, which depicts the irradiance measured in the same place on a cloudy day, whereby that nice bell curve collapses unpredictably. Zooming into the 11:00 am peak period also highlights the substantial difference in irradiance even among different sections of the array (each colored line denotes measurements taken by separate pyranometers throughout the solar farm). It is easy to see from these charts how dynamic and complex this variability can be.

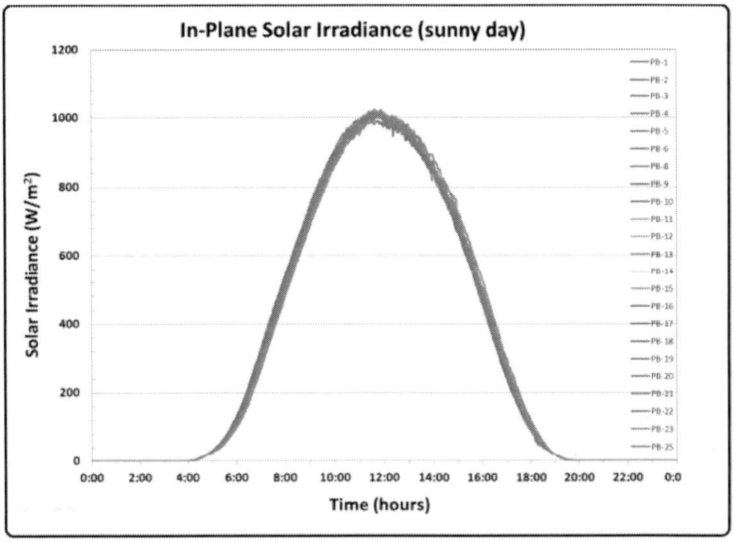

Figure 13. Amount of sunlight received by LISF on a sunny day.

This variability is a fundamental problem for the PV solar industry. It is important to note, though, that the steep ramp rates and unpredictability of solar power do not have a dramatic impact on the grid. This is because solar energy represents only a minute amount of power compared to the rest of the full energy mix. The LISF, for example, contributes only about 1% of the power supply for Long Island (Deering 2012). This should not diminish the importance of this issue, though.

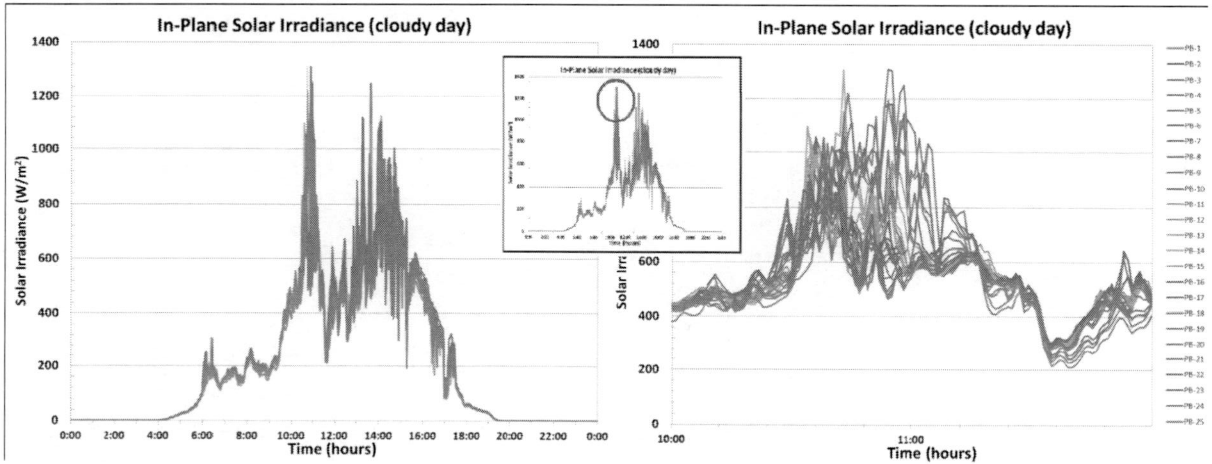

Figure 14. Amount of sunlight measured at LISF on a cloudy day.

In order for solar energy (and other highly variable resources) to increase its percentage contribution to the overall generation portfolio, its predictability and reliability must be dramatically improved. This poses a cutting-edge problem that the LISF project is well- equipped to help solve.

The major goals of CRADA 1's key areas include the following (BNL Solar Research Agenda 2010):

1) Variability and nondispatchability:
   - Develop database with time-resolved solar resource and meteorological data for BNL (Northeast United States) and solar plant electrical performance
   - Study the effects of weather variations and cloud transients on large-scale system output
   - Develop models for forecasting of available solar irradiation and impact on power production (minutes, hours, day ahead)
   - Examine relationship between solar radiation input and solar plant output
   - Participate in the NREL PV Variability Database.
2) Grid Integration of PV solar generation:
   - Establish the performance characteristics of large-scale solar PV plants located in the Northeast and factors affecting capacity credit
   - Evaluate impacts of large-scale solar PV arrays on the reliability and stability of the grid and local distribution systems
   - Characterize the response of various system elements (e.g., inverter recovery) following rapid changes in solar insolation due to cloud transients and how it impacts array output
   - Develop models to predict the response of the solar plant and individual components (e.g., inverters) to fault conditions
   - Evaluate the impact of surface soiling on solar plant performance
   - Study long-term performance and reliability of system and components
   - Firm LIPA's capacity credit for the LISF.
3) Environmental and ecological impacts of large-scale PV solar arrays:
   - Perform life-cycle assessments of large-scale solar PV plants and energy policy options

- Study the effect of the array to the local environment, including microclimate, meteorology, and hydrology, and the associated impacts to the habitats for local plants and animals
- Collect data useful for evaluating the impacts of large-scale solar plants to regional and global climate, including greenhouse gas emissions
- Observe and document impacts to local plant and animal life near the large array.

CRADA 2: The Northeast Solar Energy Research Center (NSERC) Overview

The second part of the research agenda is the development of the dedicated research array, the NSERC. As shown in Figure 15, this is a completely separate and much smaller array than the solar farm. The NSERC will be used as a testing ground for all the different component parts of solar power technology. This 1 MW array would generate power that would remain behind the meter at Brookhaven National Laboratory, helping the lab move toward its internal sustainability goals (Lofaro 2012). As of this publication, the land has been cleared for development, but the NSERC is still in its design phase. Construction is set to take place in the fall, and the NSERC is expected to be open for research in early 2013.

Modular Design

The concept of the NSERC makes it a unique proving ground for solar energy research and technology. The entire research center will be separated into three "solar array areas," each of which will be configured for different research functionalities. Generally those are:

- Area 1: Testing inverters and other electronic components (approximately 800 kW)
- Area 2: Testing tracking technologies and structural engineering (approximately 100 kW)
- Area 3: Testing new PV module technologies (approximately 100 kW)

Figure 15. Location of the Northeast Solar Energy Research Center.

Each of these arrays will have a modular design to the greatest extent possible. This will enable users to test innovations in various component parts of the system. The "plug and play" concept adds a great deal of technical complexity to the design and naturally increases the overall design costs. What it will provide, though, is an extremely valuable opportunity. It will be a unique place for industry to come and test new solar technologies. Often times it is prohibitively difficult for companies or individuals to get through the permitting process just to test new ideas. The NSERC could be that perfect incubator test bed for all sorts of new ideas to develop (Chandler 2012). The NSERC will set those conditions for industry. It will enable researchers to prove new technologies and conduct crucial research to improve the efficiency and deployability of PV solar in the Eastern United States, and more broadly.

No Specific Agenda

It is important to realize that the research component of the solar farm project is not a research project unto itself, but rather it is the development of a laboratory facility to conduct a myriad of research projects *for* industry, *by* industry. Brookhaven does not have particular research that it wants to conduct for its own purposes. BNL's intention is to establish a resource for industry to conduct cutting edge research and development for solar energy (Lofaro 2012). Below is a sample menu of research areas that NSERC is intended to support:

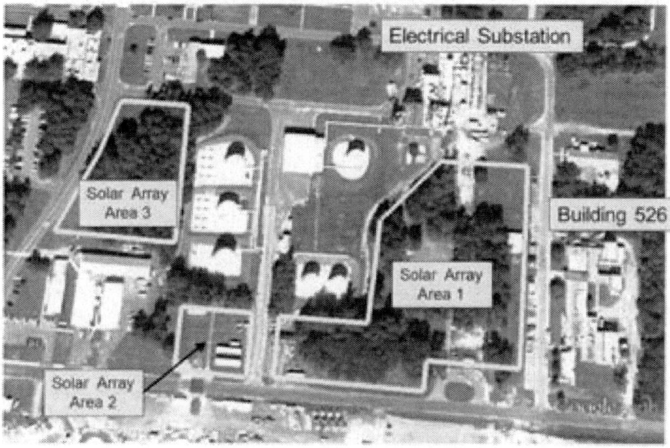

Figure 16. Development Plan for the NSERC Arrays.

Figure 17. NSERC Array plan.

Solar Energy Research: Standardized testing of solar system components:

- PV panel technology
- Inverters and micro-inverters
- Converters and micro-converters
- Advanced metering infrastructure

Storage Systems Integration

- Batteries
- Mechanical storage
- Chemical storage

Smart Grid Technology Applications and Study

- Modeling
- Advanced metering infrastructure technologies

Solar Simulators and Forecasting Models

- Near term (minutes)
- Midterm (hours)
- Long term (next day/s)

Load Simulators
Failure Analyses and Vulnerability Mitigation
National PV Environmental Research Center

- Environmental, health, and safety aspects of photovoltaic systems
- Life-cycle analyses

Meteorological Services Program

- On-site met services

RESULTS

Though the project is relatively new, and in many ways has only just begun, there are many different ways to measure the results coming out of the Long Island Solar Farm. As of this writing, the NSERC has yet to be built, so it has yet to produce results [other than the fact that it has generated a lot of attention from the PV solar industry and energy research groups (Lofaro, Looney 2012)]. The large array has been up and running since November 2011, and the preliminary results already indicate resounding success. The project as a whole has been remarkably beneficial for each of the parties involved.

The Public

All said and done, this project presents real value for a reasonable price. Michael Hervey, chief operating officer of LIPA, explains that, "[The cost amounts to] about $0.60 per month for the typical residential customer on Long Island for the next 20 years. In addition, (the power purchase agreement with the LISF, LLC) not only allows us to provide clean energy to LIPA's customers, but it also delivers price stability for our customers in an energy market where oil and gas prices remain volatile" (CCFE, 2012). Though PV solar energy remains comparatively expensive and the financial costs are shouldered by Long Island ratepayers, that price is generally considered affordable compared to the perceived benefits received by the community.

Environmental Groups

The project is a success when considering the opportunity costs to the environment from generating an equivalent amount of electricity from a fossil fuel source. Through that lens, the LISF results in the abatement of more than 30,000 metric tons of carbon dioxide emissions per year, as well as substantial amounts of other pollutants such as nitrogen oxide (NOx) and sulfur dioxide. These environmental benefits, combined with a thoughtful and comprehensive stakeholder engagement approach, helped the project earn the Best Photovoltaic Project of Year Award (for 2011) from the New York Solar Energy Industries Association (NYSEIA).

Though the project had a substantial impact on the environment on federal property, DOE, BP Solar, and LIPA made a determined effort to engage with environmental groups, namely the Pine Barrens Society, the Long Island Native Plant Initiative, and the Pine Barrens Commission. The result was the natural resource benefits package, which was a robust suite of mitigation efforts that left environmental groups largely satisfied.

U.S. Department of Energy

The project is a success for DOE, who was able to create a win-win-win scenario whereby the local community benefits, the business sees a profit, and the whole project is in keeping with the Department of Energy's mission. It also demonstrated that a project like this can be done where it was traditionally thought it would be impractical or even impossible: on a Superfund site, and in sensitive environmental conditions, and in the Northeast in general. It accomplished all of this in an innovative way and in keeping with DOE's mission. Members of the DOE at Brookhaven were also awarded Secretarial Achievement Awards for their work on this project.

Atmospheric Impacts

- CO_2 Avoided: 30,950 metric tons/year (1.2 million metric tons over 40 years) Compared to conventional electric generating facilities on Long Island
- CO_2 Sequestration Lost: 842 metric tons/year lost due to removal of trees (33,680 metric tons over 40 years)
- Methane Avoided: 80 metric tons over 40 years

Environmental Benefit Contributions

- BNL formally protecting 89 acres (51 + 38) in site master plan
- LIPA formally protecting 45 acres
- LISF contributing $75,000 to LI Native Plant Initiative for environmental restoration
- $2 million of forgone consideration to preserve additional property

Other Pollutants Avoided

- 1,320 metric tons of Nitrogen Oxides (NO_x) avoided over 40 years
- 11,200 metric tons of Ozone Season NO_x avoided over 40 years
- 3,040 metric tons of Sulfur Dioxide avoided over 40 years
- The energy required to produce and construct arrays equals approximately 2 years of output (88,000 MWh equivalent)

Brookhaven Science Associates, Brookhaven National Laboratory

The results for Brookhaven National Laboratory have been very positive, and are only just beginning. The unprecedented amount of instrumentation that has been built into the solar farm is already producing huge amounts of data. This is providing uniquely valuable data sets for researchers to analyze. The data is driving innovations in solar forecasting models, large-scale grid integration, and environmental effects, among other areas of research.

At the time of this writing, BNL is still working with BP Solar to develop the NSERC. Once that is operational (projected to be in early 2013), BNL will be able to open its doors to industry as the premier solar research and development facility in the east.

Long Island Power Authority (LIPA)

According to Mike Deering, this is a very successful project and the array is producing power very well. The best results are that, with this one project, LIPA was able to:

- Bring in 32 MW of clean electricity to the grid
- Avoid using fossil energy as fuel with its fluctuating costs and environmental hazards
- Significantly reduce peak energy load
- Diversify LIPA's portfolio with renewable resources on Long Island
- Improve economic development; at the time of peak construction, it was the largest construction project on Long Island
- Generate renewable energy credits toward LIPA's renewable portfolio standard.

Mr. Deering also stated that, "Long Islanders are proud of this project. Accolades for it have come from every diverse corner of Long Island society and interest groups—Businesses, Environmentalists, Democrats, Republicans, and everyone in between" (Deering 2012).

BP Solar

In December 2011—the very next month after the LSIF became operational—BP Solar declared it would be going out of business. This was a financial decision for the broader company based on macro-level decisions, which were a long time in the making. The Long Island Solar Farm did not contribute to that decision; it was a profitable project for BP Solar (Chandler 2012).

According to Richard Chandler, "BP Solar really built a platinum project at BNL with a long- term trajectory in mind. It's a great facility, with a strong partner in DOE and LIPA, and it has a strong balance sheet; all the risk has been taken out. Plus, by the numbers, the array is performing well."

He added that the best aspect about this project is that it's not just another power array. It has increasing value over time because of the research piece. There are opportunities to get great information from the research being done on the large array, and the NSERC is an innovation that will surely create more public-private partnerships to move the industry forward.

Implications and Next Steps

While some of the results of this project are evidently off to a successful start, the real benefits of the Long Island Solar Farm and the NSERC have only just begun. This endeavor is replete with new implications for each of the parties involved.

The Public and Environmental Groups

This project has afforded the community and environmental groups several benefits with regard to generating clean energy, avoiding polluting alternatives, and mitigating its environmental impacts. This project sets a precedent in two overarching ways.

First, this project represents a new opportunity for development. With the establishment of a successful commercial project on federal lands, it is likely that other investors and developers will come knocking at DOE's doorstep. Depending on the types of projects that are proposed or authorized, the effects on the environment and the rest of the community will vary.

Second, this project establishes a very high standard of engagement and interaction between project developers, DOE, and the public. The proponents of the LISF project were very meticulous in their engagement with the community right from the start. They established open lines of communication among all the parties. This kept the public involved and informed, which facilitated the project moving relatively smoothly through the design, permitting, and construction processes. This strategy effectively diffused any major opposition from the public or environmental groups, which could have been easily fomented by disenfranchisement from the process.

Any new project will have to genuinely engage and involve the public in the same way. The community and environmental groups on Long Island not only deserve it, they expect it, too.

U.S. Department of Energy

As the administrator of the federal property at Brookhaven National Laboratory, DOE went out of its comfort zone to make this project a reality. DOE assumed significant risk in offering to host a commercial solar farm, exposing the department to potential backlash from the community and other federal agencies who could claim that this project was, at face value, an illicit use of public lands for private profits.

The DOE demonstrated, however, that a project like this can be done and well within the parameters of the department's mission and in the best interests of the lab. They also demonstrated the importance of executing a thorough public outreach program and ensuring the cooperation of the project developer with an up-front investment of $2.3 million from BP Solar. This insulated the DOE from the business risk that BP Solar would go out of business. Nobody expected they would, but they did. Thus, DOE's insistence on receiving that down payment saved the viability of the project for DOE and the lab from BP Solar, who would have otherwise left the project without a research array. This serves as an important template in this kind of public-private partnership.

The LISF project also very much falls in line with the DOE's Asset Revitalization Initiative, or more formally known as the Energy Parks Initiative. The EPI is designed to leverage "federal assets to increase the taxpayers' return on investment" (Gilbertson 2009). In practical terms, "EPI will convert [Office of Environmental Management] liabilities (e.g., contaminated sites, facilities, and materials) into assets to solve critical national energy issues." This includes, but is not limited to, the repurposing of federal lands for the development of renewable energy[9] (Gilbertson 2009).

In fact, according to Bob Gordon, Brookhaven National Laboratory was approach by people in the [Energy Parks Initiative]. "They were wanting to uphold [The Long Island Solar Farm] as an example of their initiative. This project was not, however, developed under that program, and it preceded the inception of the [Energy Parks Initiative] altogether."

Clearly, the LISF is in the general spirit of what the Department of Energy is trying to do in the future—leverage available federal facilities for the development of innovative energy projects. Though it is not directly associated with the Energy Parks Initiative, the Long Island Solar Farm serves as a successful example of what can be achieved.

Brookhaven Science Associates, Brookhaven National Laboratory

According to Dr. Looney, "The Long Island Solar Farm and the research array have changed the face of BNL to a certain extent. It has become a visible symbol of the future of energy technology and the Brookhaven's role at its cutting edge. The project also created a new relationship between BNL and LIPA, as well as new relationships with DOE/EERE program managers."

BNL has already been approached by EPRI, who wanted to join us in their solar energy research. National Center for Atmospheric Research (NCAR) also came to us for test-bed data as it relates to solar energy. These are two very clear examples of how major organizations in the PV solar industry recognize the research and development opportunities at Brookhaven (Looney).

Long Island Power Authority (LIPA)

There were some key lessons that LIPA took away from this project, particularly with regard to its experience with the request for proposals. The solar RFP was the first of its kind for LIPA and New York, and LIPA soon discovered that the RFP process was very time and labor intensive. Though LIPA was no stranger to RFPs, this one attracted a great deal of attention. As a result, the whole RFP process took quite a lot of time and staff-work to prepare the request, submit it, field questions, and organize and analyze all the responses. That was all before routing recommendations through LIPA's chain of authorities, which was time consuming in itself.

Another drawback of the RFP process was that, in the public's eye, LIPA ostensibly became associated with the "developers" of the project. The nature of the RFP and the review procedures required LIPA to be very much engaged in the process, more so than they would have preferred to be. Such involvement gave outsiders the general impression that LIPA was a principal developer of the project, when that was not the case (Deering). LIPA's primary interest was the power resource, not any one particular style or program of development.

From those two major lessons, LIPA has moved towards using feed-in tariffs (FITs) as the primary vehicle for connecting PV solar projects to the grid. Instead of outlining individualized parameters solar power projects, LIPA has chosen to standardize the policies by which solar projects, big and small, can connect to the grid. This establishes clarity and certainty for developers and helps LIPA avoid being in the weeds of the development process (Deering 2012).

LIPA generally expects to see more distributed PV solar programs develop across Long Island. It is unlikely that Long Island will see another PV solar array built on a scale close to the LISF, but there is still a lot of potential for double-digit megawatt arrays (Deering 2012).

BP Solar and the PV Solar Industry

BP Solar is committed by a $2.3 million down payment to finishing the development of the NSERC before it goes out of business and disengages completely from working with the lab. Their follow-through with this consideration aspect of the easement agreement will not necessarily benefit BP Solar directly, but it certainly will help set conditions to advance the PV solar industry through research and innovations made at the NSERC.

The implication for investors and developers in the broader PV industry are two-fold. First, the Long Island Solar Farm is a new source of truly vast amounts of information and research capabilities that are unavailable elsewhere, certainly not in the Eastern United States. Second, this project represents an example of an effective public-private partnership that is mutually beneficial and profitable as a commercial enterprise. This project opens new possibilities for industry to work with federal and state agencies in places where there may be the presumption that those organizations would be uninterested in this type of PV solar deployment. Where there is a will, there certainly is a way.

CONCLUSION

Is the Long Island Solar Farm
a good template
for utility-scale PV solar development in the
Eastern United States?

Not a Good Template

According to the Oxford dictionary, the definition of a "template" is "something that serves as a pattern from which other similar things can be made." If we apply this definition to the Long Island Solar Farm it certainly presents a compelling case study as an example of success for a variety of different stakeholders. There were, however, some remarkably nuanced conditions that enabled this project to become a reality. It is the uniqueness of those circumstances and conditions that prevent this project from being imitated in many places, if anywhere else at all.

Some of those unique conditions include:

List A: Unique Circumstances

- No transmission was necessary; LIPA had an existing substation immediately adjacent to the proposed site
- The availability of very large tracts of unused land on federal property in close proximity to load
- The classification of the lab property as a "Superfund site" created a voluminous body of thorough environmental data
- The lab's history as a military installation generated a trove of detailed surveys of historical and cultural artifacts on the property
- The mission of the federal facility was such that a commercial project could be made mutually beneficial

There were also a number of conditions that were absent from the LISF site, any one of which could have derailed the LISF project or prevented it from happening altogether. Some of these include:

List B: Anti-Conditions

- Endangered or protected species of plants or animals (although the endangered tiger salamander is widely prevalent on the site, there was space enough for the project to be built around its protected habitat)
- Legal restrictions on land use within the federal enclave
- An irreconcilable opposition group

Clearly, the conditions in List A represent a combination of very fortunate circumstances that greatly enabled the development of the Solar Farm. If any single one of those conditions did not exist, for example, it could have had dramatic impacts on the time it would take to develop the project, the overall costs, or both. Additionally, the "anti-conditions" in List B are circumstances that could easily be found in many other locations and, if present, would likely prevent a project like this from developing.

The aggregate of these conditions makes the experience of the Long Island Solar Farm very difficult, if not impossible, to replicate elsewhere. Therefore, even though the Long Island Solar Farm has proven to be a success story for large-scale PV solar energy, it does not readily "serve as a pattern from which other similar things can be made." For that reason, this project is not a good template for utility-scale solar development in the East.

A Trailblazing Resource

The Long Island Solar Farm does not necessarily have to be replicable to serve as an example. Instead of *being* a good model *of* development, the LISF provides a unique opportunity to *build* good models *for* development.

The research agenda at Brookhaven National Laboratory, including the vast instrumentation on the solar farm and the dedicated NSERC array, will provide essential data and research opportunities. The instrumentation on the solar farm is already generating valuable information that is supplying the development of better solar forecasting models.

The NSERC will also give industry an unparalleled opportunity to push the limits of PV solar technologies and test their applications in the East. The research agenda at BNL will greatly enhance industry's understanding of the conditions they face in building PV solar installations in the region. By answering some of the major questions for industry, the Long Island Solar Farm will thereby reduce the barriers to future development of major solar projects in the east. So, in the end, even though it may be impossible to literally follow in the LISF's footsteps, industry will certainly be able to find its way down a clearer path forward, thanks to the research being done at the lab.

A Good Example: Innovation in Attitude

The Long Island Solar Farm is certainly a remarkable success story in that it has become a large-scale and mutually beneficial asset for a diverse group of public and private stakeholders. What makes the LISF story so exemplary is that, in all likelihood, it really never should have happened in the first place. As Mike Deering described, the group that came together represented "the right organizations, the right people, and the right attitude to see it through." Dr. Looney agreed that the group was "a real coalition of the willing."

Table 2 shows how unlikely this project was from the start. Each of the proponents chose to get involved and commit to finding success. Nobody would have faulted any of these groups had they remained with their most likely course of action, which was to deny the project.

The table demonstrates how the participation of each of these groups was a distinct departure from what was expected of them. If one of these groups had chosen to adhere to their most likely position, the project would not have gotten off the ground.

In addition to mass—and unlikely—participation, a distillation of the Table 2 down to its essential parts reveals some key conditions and decisions that were utterly crucial to development.

List C: Key Conditions and Decisions

- DOE's willingness to host a commercial venture on federal property
- LIPA's selection of BP Solar's radical proposal
- BP Solar's willingness to sign an easement instead of a lease
- BP Solar's willingness to use consideration as a means to fulfill DOE mission requirements
- DOE's willingness to legally enable BP Solar to cash-out of consideration obligations (separate from the down payment on the NSERC)
- A joint effort from LIPA, BP Solar, DOE, and BSA on a robust natural resources benefits package
- A vigorous communication and outreach plan with the public
- Respectful inclusion of all interest groups, even (and especially) those in opposition

Each of the parties involved in the development of the LISF recognized the unusual opportunities that the favorable conditions in Lists A and B presented. This is what helped pull each group from their most likely positions. Furthermore, this helped buttress a sense of teamwork among the disparate groups, which enhanced their collective ability to make decisions and set the conditions to ensure successful completion of the project.

Table 2. Finding Success in Unlikely Positions

	Most Likely Position (No!)	**Actual Position (Yes!)**	**Risk**
LIPA	1) Requesting small-scale, distributed solar projects 2) Rejecting a radical proposal from BP Solar	1) Opening the possibility of a utility-scale project with a bold solar RFP 2) Selecting BP Solar's proposal	1) All eggs in one basket 2) A commercial project on federal land would languish in permitting and public opposition

	Most Likely Position (No!)	Actual Position (Yes!)	Risk
BP Solar	1) The company only develops smaller projects (10-15 MW) 2) Can only tolerate a lease agreement 3) BP Solar is a business; not interested in supporting a research agenda	1) Proposed a 37 MW array 2) Agreed to an easement 3) Using consideration could add value to BP Solar and support the DOE mission	1) Unfamiliar territory for BP business models, especially with variable conditions in the East 2) Loss of long-term land control 3) Financial loss
DOE	Potential conflict of interests; a federal agency cannot support a commercial development project on public lands	DOE can support a commercial venture if the project fits the mission of the Department of Energy	Public backlash in support of the most likely position
BNL	A commercial project will not fully support a research agenda	Allocating use of consideration to develop research agendas; insured with down payment	Embarrassment for the Lab if research agenda disintegrates or never materializes
Public	DOE and BNL are outside of their purview; should not support commercial development at the national lab	If done properly, the project would benefit the community	Setting a precedent for commercial development at BNL or on other public lands
Environmental Groups	This project comes at a high cost of crucial Pine Barren ecosystems; could incite firestorm of public opposition	Mitigation efforts could be effective to compensate for forested lands being cut down	Setting a precedent for commercial development at a high cost of environmental resources

On Leadership

Leadership comes in different forms, of course. In many cases, the most efficient type of leadership results from a single decision-maker. In contrast, the experience of the LISF project does not appear to support that concept.

The development process does not distinguish any single point of leadership that was ultimately responsible for driving this project to completion. You cannot say, for instance, that DOE took the lead on this project because in many ways LIPA did, and so did BP Solar. Even the public and community leaders got involved to shape the debates and outcomes. For example, members of each group cited the work of Congressman Bishop, who was actively engaged and served as a mediator in the development debates (Looney, Gordon, and Deering 2012). While that was important and helpful, Congressman Bishop's participation was not necessarily crucial to the project's success. In the end, though, no group established itself as dominant over the other. Accordingly, on the surface, there appears to be a conspicuous absence of any single point of leadership in this endeavor.

While no single leader can be rightly identified, each stakeholder group certainly exhibited a great deal of leadership within their own camps.

Key features of this internal leadership include:

List D: Leadership Attributes

- Having an ambitious vision for the future
- Recognizing opportunities
- Taking bold action while mitigating risks
- Looking for constructive ways to achieve a positive goal versus looking for reasons not to
- A genuine respect for disparate interests
- The ability to compromise and cooperate

What is striking about the development of the Long Island Solar Farm is that each of the stakeholder groups consistently displayed every one of these essential leadership qualities. Consider for a moment if any of these qualities were absent from any of the stakeholder groups—the effects would have been disastrous for the group effort. In the same vein, if any one group tried to establish itself as the single dominant leader in this consortium of interest groups, it is likely that the project would have failed. The interests that brought each party to the table would have taken on a perception that they were there to support a single group's agenda. This would have diluted the attraction of mutual benefits that each group expected from the outcomes, and the project would have likely collapsed on itself before even getting started.

In Closing

Undoubtedly, leadership played a quintessential role in the development of the Long Island Solar Farm. Each group indeed had the "right people" who could effectively lead their organizations according to those critical tenets outlined in List D. Those leaders were able to come together to strike a balance of cooperation that structured the debates, safeguarded individual interests, and found a way to make an unlikely dream a reality. That is the essence of the Long Island Solar Farm, and that is what makes it an excellent example for public-private partnerships and the future of solar energy.

Figure 18. Leadership Star Diagram.

References

"2011 SEPA Utility Solar Rankings: SEPA's Fifth Annual Report Highlighting Utility Solar Leaders." (2012). *Solar Energy Power Association.* Accessed July 2, 2012: http://www.solarelectricpower.org/solar-tools/sepa-utility-solar-rankings.aspx.

Amper, Richard. (July 13, 2012). Long Island Pine Barrens Society, Riverhead, New York.

Bergin, Tom & Sarah, Young (Dec. 21, 2011). "BP Turns Out Lights at Solar Business." *Reuters Online.* Accessed July 10, 2012: http://www.reuters.com/article/2011/12/21/us-bp-solar- idUSTRE7BK1CC20111221.

"BNL Solar Research Agenda: Brookhaven National Laboratory Research Agenda on Solar Photovoltaic Systems." (Sept. 15, 2010). Upton, New York: Brookhaven National Laboratory.

"BP Alternative Energy: Solar Power." (2013). BP Solar. Accessed in 2012: http://www.bp.com/sectiongenericarticle.do?categoryId=9025019&contentId=7046515.

"Brookhaven Science Associates: Laboratory Administration." (2013). Brookhaven National Laboratory. http://www.bnl.gov/bnlweb/admin/bsa.asp.

"Central Pine Barrens Joint Planning and Policy Commission." (2013). Central Pine Barrens Joint Planning and Policy Commission. http://www.pb.state.ny.us/index.htm#Act_Plan_Commission.

Chandler, Richard. (July 16, 2012). Long Island Solar Farm, BP Solar International, Inc., Brookhaven National Laboratory, Upton, New York.

"Citizens Campaign for the Environment: About Us/Contact." (2012). Citizens Campaign for the Environment. http://www.citizenscampaign.org/about.asp.

"Database of State Incentives for Renewables & Efficiency." (2012). DSIRE. Accessed July 13, 2012: http://www.dsireusa.org.

Deering, Michael J. (July 26, 2012). Long Island Power Authority, LIPA Headquarters, Uniondale, New York.

EA-1663. "Environmental Assessment for BP Solar Array Project." (December 2009). Brookhaven Site Office, Office of Science, U.S. Department of Energy.

Easement Outgrant. "For Installation and Operation of a Solar Electric Generating System at Brookhaven National Laboratory." (January 2010). Agreement between the United States Department of Energy and the Long Island Solar Farm, LLC.

Easement MOU. "Easement Memorandum of Understanding for Cooperation in Solar Energy Between BP Solar International, Inc. and the United States Department of Energy." (February 2010). Attachment 15 to the Easement Outgrant between the United States Department of Energy and the Long Island Solar Farm, LLC.

"Environmental Considerations Map." (2010). *Research Agenda on Solar Photovoltaic Systems.* Upton, New York: Brookhaven National Laboratory.

"EPA Region 2 Superfund: Brookhaven National Laboratory." (2011). Environmental Protection Agency. Accessed Aug. 8, 2012: http://www.epa.gov/region2/superfund/npl/brookhaven/.

EPIA. (May 2012). "Global Market Outlook for Photovoltaics Until 2016." *European Photovoltaic Industry Association.* Accessed July 10, 2012: http://files.epia.org/files/Global- Market-Outlook-2016.pdf.

FERC. (May 2011). "Other Markets: Renewables & Energy Efficiency—Generation & Efficiency Standards." *Federal Energy Regulatory Commission Online.* Accessed Aug. 8, 2012: http://www.ferc.gov/market-oversight/othr-mkts/renew.asp.

Geary, P. (May 25, 2005). "Solar Power Profitability: BP Solar." *Environmental News Network.* Accessed July 10, 2012: http://www.climateark.org/shared/reader/welcome.aspx?linkid=42124.

Gilbertson, M. A. (April 23, 2009). "Energy Parks Initiative." The Office of the Deputy Assistant Secretary for Engineering & Technology, U.S. Department of Energy. Accessed July 14, 2012: http://www.em.doe.gov/pdfs/Gilbertson_EnergyParkInitiativeUpdate.pdf.

Gordon, R. P. (June 27, 2012). Brookhaven Site Office, Office of Science, U.S. Department of Energy. Brookhaven National Laboratory, Upton, New York.

Green, T. M. (July 17, 2012). Environmental Protection Division, Brookhaven Science Associates. Brookhaven National Laboratory, Upton, New York.

Lenardic, D. (July 13, 2012). "Large-Scale Photovoltaic Power Plants, Ranking 1-50." *PV Resources.* Accessed July 18, 2012: http://www.pvresources.com/PVPowerPlants/Top50.aspx.

"LIPA, BP Solar and Brookhaven National Lab Flip the Switch at the Long Island Solar Farm." Long Island Power Authority. Accessed July 11, 2012: http://www.lipower.org/newscenter/pr/2011/111811-solar.html.

"LIPA Chooses BP, enXco for Solar Installations." (March 19, 2012). *Sustainable Business.com.* Accessed July 10, 2012: http://www.sustainablebusiness.com/index.cfm/go/news.display/id/17827.

"LIPA: Company Profile." (2013). Long Island Power Authority. http://www.lipower.org/company/profile/.

"LIPA Electric Resource Plan 2010-2020." (February 2010). Long Island Power Authority. Accessed July 10, 2012: http://www.lipower.org/company/powering/ energyplan10.html.

"LIPA Energy Plan 2004-2013, Volume 2. Energy Primer." (June 23, 2004). Long Island Power Authority. http://www.lipower.org/pdfs/company/projects/v1.energyplan.pdf.

Lofaro, R. J. (July 10, 2012). Renewable Energy Group, Sustainable Energy Technologies Department, Brookhaven Science Associates. Brookhaven National Laboratory, Upton, New York.

"Long Island Solar Farm Largest in State History." (Nov. 21, 2011). Citizens Campaign for the Environment. Accessed July 13, 2012: http://www.citizenscampaign.org/news/story.asp?id=465.

Looney, P. J. (July 17, 2012). Sustainable Energy Technologies Department, Brookhaven Science Associates. Brookhaven National Laboratory, Upton, New York.

"New York State Department of Environmental Conservation: Eastern Tiger Salamander Fact Sheet." Accessed Aug. 6, 2012: http://www.dec.ny.gov/animals/7143.html.

"NREL: Dynamic Maps, GIS Data, & Analysis Tools." National Renewable Energy Laboratory, Golden, CO. Accessed July 10, 2012: http://www.nrel.gov/gis/solar.html.

Ryan, M. (Aug. 31, 2011). "Tax Break Tactics Divide Renewable Industry." *AOL Energy*. Accessed July 10, 2012: http://energy.aol.com/2011/08/31/tax-break-tactics-divide-renewable- industry/.

Sadler, L. (July 19, 2012). Brookhaven Site Office, Office of Science, U.S. Department of Energy. Brookhaven National Laboratory, Upton, New York.

"SERDP/ESTCP: Energy and Water." Strategic Environmental Research and Development Program (SERDP) and Environmental Security Technology Certification Program (ESTCP). Accessed July 5, 2012: http://www.serdp-estcp.org/Program-Areas/Energy-and-Water.

"SMUD's PV1: Solar Power for 25 Years." (Sept. 9, 2009). *Green Energy News*. Accessed July 19, 2012: http://www.green-energy-news.com/nwslnks/clips909/sep09011.html.

Snook, J. (March 7, 2007). "Department of Defense Completes Renewable Energy Assessment." U.S. Department of Energy—Energy Efficiency and Renewable Energy, Federal Energy Management Program. Accessed July 12, 2012: http://www1.eere.energy.gov/femp/news/printable_versions/news_detail.html?news_id=10612

"The Long Island Pine Barrens Society." Long Island Pine Barren Society. http://pinebarrens.org.

"United States Department of Energy: Office of Science." (2013). U.S. Department of Energy. http://science.energy.gov/laboratories/site-offices/.

Wiegand, P. (2011). "Long Island Native Plant Initiative." Friends of the Bay. Retrieved on July 21, 2012 from: http://friendsofthebay.org/?p=1200.

Wood, E. (June 20, 2012). "Solar in New York: Its Strategy to Make Solar Shine" *Renewable Energy World.Com.* Accessed July 19, 2012:

http://www.renewableenergyworld.com/rea/news/article/2012/06/solar-in-new-york-its-strategy- to-make-solar-shine.

End Notes

[1] This statistic was taken at the time of its commissioning in November 2011. As of July 2012, the LISF is tied for 28th in the world due to larger projects being constructed in India, Europe, and the Western United States (Lenardic 2012).

[2] This statistic is from the time of its commissioning in November 2011. As of July 2012, the LISF is ranked the 7th largest in the country due to very large projects that have come online in Arizona and Nevada (Lenardic 2012).

[3] The DOD is currently at approximately 10% toward a long-range goal of 25% by 2025 (SERDP 2012).

[4] Starting at 19% in 2004 (DSIRE 2012).

[5] DOE informed its use of an easement, in part, by a similar land-use arrangement made for a PV solar developer on federal property administered by DOE at the National Renewable Energy Laboratory in Colorado (Sadler 2012).

[6] LIPA's original breakdown of the 50 MW had 37 MW coming from BP Solar and 13 MW from enXco. LIPA reallocated 5 MW from BP Solar to enXco during the design phase of both projects, leaving 32 MW for the LISF (Deering 2012).

[7] See the "Research Opportunities" section for more details.

[8] See the section on "Environmental Impacts" for more details.

[9] Other EPI uses for federal lands cited include: renewable energy (solar, wind, biomass, geothermal); fossil fuels (clean coal, gas turbines); electricity generation, transmission, distribution; hydrogen generation; emission controls (carbon sequestration, specialty manufacturing); and nuclear (power, fuel cycle, waste management).

In: The Long Island Solar Farm
Editor: Clara Mckenzie
ISBN: 978-1-63117-419-3

Chapter 2

ENVIRONMENTAL ASSESSMENT FOR BP SOLAR ARRAY PROJECT [AT] BROOKHAVEN NATIONAL LABORATORY*

U.S. Department of Energy, Office of Science and Brookhaven Site Office

1.0. SUMMARY

BNL is a national laboratory overseen and primarily funded by the Office of Science (SC) of the DOE, and operated and managed by Brookhaven Science Associates (BSA), a limited liability company founded by Stony Brook University and Battelle, a nonprofit, applied science and technology organization. Located 60 miles east of New York City in Upton, NY, BNL conducts research in high energy and nuclear physics, chemistry, nanotechnology, environmental sciences, energy technologies and national security (See Figures 1 and 2). Among its missions, the Laboratory is charged with conceiving, designing, constructing and operating world-class, complex, leading-edge research facilities in response to the mission needs of DOE and to a large community consisting of university, industry, government and international users [SER].

* This is an edited, reformatted and augmented version of U.S. Department of Energy publication DOE/EA-1663, dated December 2009.

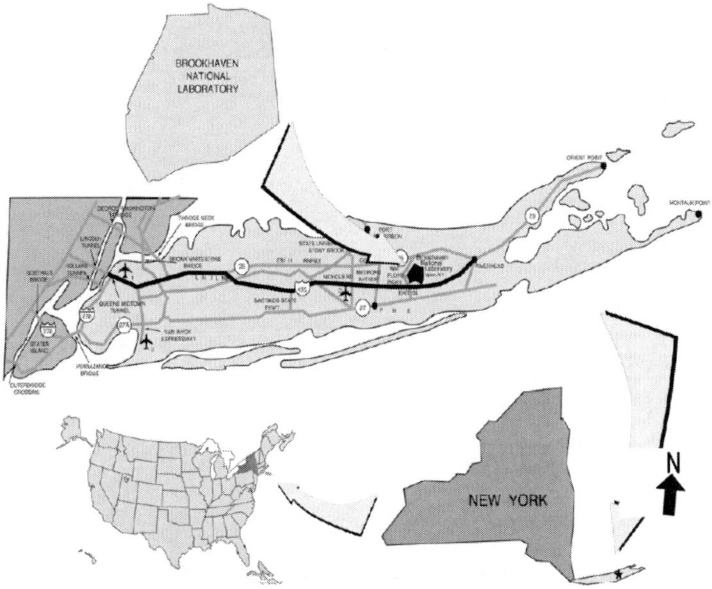

Figure 1. Regional View of Brookhaven National Laboratory Location.

Figure 2. Aerial View of Brookhaven National Laboratory Core Developed Area.

If constructed, the proposed BP Solar Project would be located on approximately 200 acres and would encompass two large-scale commercial solar photovoltaic (PV) arrays totaling approximately 37 MW, and potentially a smaller Laboratory dedicated array of 1 to 2 MW, with an associated support building. This project grew out of solicitation by the Long Island Power Authority (LIPA) to diversify its portfolio with renewable energy sources. LIPA publically solicited companies for proposals to provide solar-generated power on Long Island. In that solicitation, DOE offered BNL property as a host site for a project that would meet LIPA's needs. BP Solar's successful proposal to LIPA for a 37 MW photovoltaic system on the federal property resulted in negotiations between DOE and BP Solar for access to the site for construction and operation of the system. The project supports DOE's renewable portfolio and the Department's vital goals of reducing dependence on fossil fuels and encouraging research and development in energy technologies. Additionally, this proposed solar project is consistent with NY State's goal to meet 45 percent of its electricity needs through improved energy efficiency and renewable sources by the year 2015.

The proposed solar project at BNL promotes the use of non-polluting, renewable solar energy technology in the largest-scale photovoltaic project in the Northeast. Power generated by this array will be placed into the LIPA grid and is expected to help LIPA meet its ever-increasing electric demand. Because of the size of the proposed project, it presents an unprecedented opportunity to obtain data and practical experience in areas such as power intermittency effects on the utility grid, technological and material efficiency, power storage and climate. This data would be vital to supporting the business case for the proliferation of large solar arrays across the U.S. The data obtained from this array could also be vital to the research and development efforts designed to improve solar technologies and impacts on large utility electric grids. In addition, the project would provide the opportunity to explore new research collaborations and launch educational and community outreach programs on solar energy as a renewable energy source. As partial consideration for use of the federal property, BP Solar may sponsor Work for Others projects at BNL that would allow BSA to perform research or other technology work for BP Solar, a non-DOE entity, on a fully reimbursable basis.

While no specific upgrades to the solar arrays or project footprint are planned at this time, it would be reasonable to expect that technological advances may warrant changing or modifying some of the solar arrays or its associated equipment in the future to achieve improved efficiency. At the end of operational life, estimated to be about 40 years, the arrays would be dismantled and disposed of as part of a planned decommissioning process. All support structures and associated electrical hardware (wiring, conduit, towers, etc.) would also be removed. The land would then be available for future BNL operations, restoration, or remain open to natural re-growth.

Summary of BP Solar Project estimated parameters:

- Total generation capability of 37 MW direct current (DC) - Solar arrays would consist of approximately 167,712 PV modules, each about 5.5 x 3.3 x 0.016 feet (1,667 x 1,000 x 50mm), weighing 43 pounds (lbs) (19.4kg), and having a standard rating of 220 watts.
- Estimated annual output of 1,200 Megawatt hours (MWH) per MW PV, for a total of 44,400 MHW per year.
- Two-500 kilowatt (kW) inverters and a 1-mega volt-amps (MVA) 200/13.8 kilovolt (kV) transformer would be located on each of thirty two 270-square foot (25 square meters) concrete equipment pads located throughout the arrays.
- A 15kV outdoor, metal-clad, switchgear, which is to be shelter-aisle or enclosed in an equipment building with a concrete pad of approximately 250-square foot (23 square meters).
- A 13.8/69 kV transformer, mounted on a 2,500-square foot (232 square meters) concrete pad, mounted on a pad with associated circuit breakers, disconnect switches, and a small (12' x 24') control enclosure on nearby pads all totaling 2000-square foot (186 square meters), would connect the solar array facility via transmission line to the LIPA substation located adjacent to the BNL property, south of the Long Island Railroad (LIRR) tracks.
- Fencing would enclose solar arrays and associated equipment.
- Laboratory - dedicated 1 to 2 MW solar arrays.
- Laboratory - dedicated array support building about 2,000 square feet (186 square meters).

This EA analyzes the potential environmental impacts of DOE's granting an easement to BP Solar to construct and maintain a 37 MW solar generating system and a 1 to 2 MW Laboratory dedicated array, and the No Action Alternative.

In the No Action Alternative, the BP Solar Project would not be constructed and operated at BNL. This alternative would maintain the current conditions on the BNL site. An alternative suitable location would need to be located in order to provide the local region with a photovoltaic source of electric power. If no other site exists the power anticipated to be generated from the BP solar project would need to be supplied through either conventional means, such as fossil fuel-burning power plants or imported from outside the region.

When considering potential locations for the proposed solar project, DOE and BSA utilized three primary criteria to evaluate potential sites. These criteria included 1) proximity to the LIPA substation, 2) avoidance of Core Pine Barrens areas, and 3) minimizing impacts to the environment. Additional considerations for placement included minimizing conflicts with future Laboratory development and existing infrastructure. The following alternative locations within the BNL site were initially considered and ultimately eliminated for various technical reasons: Laboratory West Boundary, Laboratory Southern Area, Dispersed Open Areas, and Rooftop Deployment. These rejected alternatives were not assessed as part of this EA.

A summary of the potential environmental impacts of the No Action Alternative and the proposed BP Solar Project is presented below in Table 1. Full analysis of these topics is covered in the Environmental Impacts section.

Table 1. Summary of Potential Environmental Impacts and Controls for the No Action Alternative and the BP Solar Project Proposed Action

Comparison Factors	No Action: BNL Current Operations	Proposed BP Solar 37 MW Project
General Information	No change from the existing BNL operations.	Construct 37 MW Solar Photovoltaic Array project: A footprint of 200 acres; 32 270-square-foot concrete pads with transformers and inverters; 15kV outdoor, metal- clad, switchgear which is to be shelter-aisle or enclosed in an equipment building with a concrete pad of approximately 250-square foot; a 13.8/69kV transformer with associated circuit breakers, disconnect switches, and a small (12' x 24') control enclosure mounted on pads totaling 2,000-square-feet; connection to existing LIPA grid at off-site substation 8ER. Construct a 1-2 MW Laboratory dedicated array and 2,000 sq. ft. support building; array potentially tied into on-site BNL electrical grid.

Table 1. (Continued)

Comparison Factors	No Action: BNL Current Operations	Proposed BP Solar 37 MW Project
Ecological Resources	Minimal expansion of invasive flora at project site could pose long-term negative impact to ecology.	*Construction Effects on Vegetation* - Disturbance of about 200 acres of land, including removal of approximately 153 acres of trees - Minimal grading and topsoil replacement would allow most understory plants to recover; native grasses or understory vegetation would also be planted. Removal of existing invasive plant species would, at minimum, slow their expansion (into adjacent wetlands, in particular). *Construction Effects on Threatened, Endangered, and Special Concern Species* – Project would meet NYSDEC protocols for tiger salamander breeding habitat. May have minimal impacts on the eastern tiger salamander beyond the 1,000 foot (304.8 meters) protection zones around known habitats. Established buffers and procedures should minimize impacts to tiger salamanders and other amphibians; temporary impacts on eastern box turtles due to construction disturbance. Fencing design may allow movement of reptiles throughout the entire area; mature forest removal would force nesting raptors to move to other suitable habitat. Overall, construction would have a minimal impact on threatened, endangered, and special concern species *Operation Effects on Threatened, Endangered, and Special Concern Species* - Planting native understory may increase numbers of host plants for rare butterflies; fencing would prevent entry of deer allowing understory vegetation to mature and seed, providing improved habitat for many threatened and endangered species. *Construction Effects on Migratory Birds* - Tree clearing and temporary understory vegetation disturbance would have a moderate effect on migratory songbirds and raptors; construction noise may

Comparison Factors	No Action: BNL Current Operations	Proposed BP Solar 37 MW Project
		disturb nesting birds; tree removal in late summer would result in some nest destruction. Timing forest clearing can minimize direct impacts; while clearing would remove available habitat, it affects 4.2% of the available habitat on the BNL site and 95.8% of available habitat remains unaffected; removal of existing fences and open areas would temporarily affect eastern blue bird nest boxes. Nest boxes can be installed on new fencing. *Operation Effects on Migratory Birds* - Slight effects on migratory birds due to increased edge habitat, improved deer-free understory, and increased fence lines for placement of nest boxes. Increased edge habitat may result in slight increases in the number of brown-headed cowbirds parasitizing songbird nests. *Construction Effects on Mammals* - Tree removal would result in mammals dispersing to surrounding forests; displaced deer would cause added stress on surrounding forests, primarily due to existing over abundance; other medium sized animals would be displaced with little effect on their populations. *Operation Effects on Mammals* - Minor positive effects on small and medium sized mammals due to fences keeping deer out of the area and improved understory. *Construction Effects on Reptiles and Amphibians* - Minimal impacts on reptiles and amphibians due to construction equipment and loss of forage and mating habitat. Minimizing understory vegetation disturbance would lessen impacts. *Operation Effects on Reptiles and Amphibians* - Slight improvements for reptile and amphibian species due to increased habitat for mole salamanders, tree frogs, snakes; maintaining habitat for turtles should result from increased understory. Proposed fencing would support movement of small animals

Table 1. (Continued)

Comparison Factors	No Action: BNL Current Operations	Proposed BP Solar 37 MW Project
		including reptiles and amphibians. *Construction Effects on Pine Barrens* - Clearing approximately 153 acres (62 hectares) of trees would have minimal impact on the overall quality of the Central Pine Barrens ecosystem; the proposed project is fully within the Compatible Growth Area on the BNL site and falls well within the 65% clearing standard established under the Central Pine Barrens Land Use Plan. *Operation Effects on Pine Barrens* - Little, if any, overall effect on the surrounding Pine Barrens; precipitation recharge to groundwater would not be affected due to use of native vegetation underneath the solar arrays and minimal impermeable surfaces.
Water Resources	No change from the existing BNL site conditions.	No expected impacts to any groundwater or surface water resource during project construction or operation. *Construction* – Storm water and silt runoff management would include silt fencing and stabilized rock construction entrances; potential modification of existing drainage and new/modified New York State Pollutant Discharge Elimination System (SPDES) outfall permit; estimated 2,000 gallons of water per day usage for dust mitigation; relocation or modification of existing water sampling wells, as necessary; Wild, Scenic and Recreational Rivers Act Permit and a Wetlands Permit, issued by the New York State Department of Environmental Conservation (NYSDEC), would be required; and a NYSDEC construction storm water permit may also be necessary. *Operation* – Soil erosion potential minimized by planting native low-growing groundcover under/around arrays; estimated 500,000 gallons of water per year for cleaning PV panels

Comparison Factors	No Action: BNL Current Operations	Proposed BP Solar 37 MW Project
		(0.1 percent of current BNL water usage). Utilize existing hydrant or former irrigation supply well; existing BNL water supply would be adequate; slight increase in impermeable surfaces (concrete equipment pads, Laboratory dedicated array Support Building) of approximately 12,890 square feet (1,178square meters), all noncontiguous; any hazardous materials storage would be managed under Suffolk County Article 12 controls. Likely modification of existing permits or acquisition of a project specific SPDES permit from the NYSDEC. While no decision has been made, pending further investigation, brownfield areas may be used for the Laboratory dedicated array and support building. If brownfield areas were used, there could be a slight increase in impermeable surfaces due to the use of ballasted systems made of concrete to support the solar panels and parking area/roadway (10,890 sq. ft.).
Land Use, Demography, and Environmental Justice	No change from the existing BNL site conditions.	Land use within the proposed project footprint would change from currently forested and open areas to groups of fenced solar arrays; no change in activities/operations performed on other areas of the BNL site; no effect on off-site land use or demographics. *Operation* - Laboratory dedicated array may result in a small increase in scientific and support staff estimated at less than 20 individuals, along with visiting research personnel. *Use of Brownfields* - Considered within industrial uses identified in BNL Land Use and Institutional Control Program. Could potentially reduce amount of forested land needed for solar arrays. *Environmental Justice* - No environmental justice impact or negative economic or health effects on any potentially affected population are anticipated.

Table 1. (Continued)

Comparison Factors	No Action: BNL Current Operations	Proposed BP Solar 37 MW Project
Socioeconomic Factors	No change from the existing BNL site conditions and operations.	*Construction* - The project would benefit construction and manufacturing sectors with secondary benefits through jobs, wages and spending. It may also provide work and training opportunities for small and emerging businesses. The estimated construction workforce would be approximately 200 full-time employees. *Operation* – The BP arrays would employ two full-time operations and maintenance personnel. The Laboratory dedicated array could employ up to 20.
Transportation	No change from the existing BNL site conditions.	*Construction* – Temporary increase in construction equipment, delivery vehicles and worker vehicles; use of alternative southeast access gate would increase vehicle traffic on local road (North Street); elimination of firebreak/access roads within project footprint. *Operation* – None.
Cultural Resources	Outhouse structure would remain at extant location. Two Camp Upton World War I (WWI) concrete foundations would remain intact.	*Construction* – Removal/relocation of outhouse structure (circa WWI era shed, relocated and converted to privy post 1921); demolition of two Camp Upton WWI concrete foundations. Architectural and archaeological surveys of outhouse site were performed; archaeological survey of Camp Upton WWI Remount Area was performed. Archaeological work was sufficient to conclude that no further archaeological investigations are recommended. *Operation* – None.
Air Quality	Reduced airborne emissions from conventional fossil fuel plants would not be realized. .	*Construction* – Temporary increase in emissions due to construction equipment, delivery vehicles and worker vehicles; generation of airborne dust ($PM_{2.5}$ and PM_{10} emissions) from grading and vehicle traffic on unpaved surfaces - dust generation controlled by water spraying on soil surfaces and stabilized rock construction entrances; minimal grading

Comparison Factors	No Action: BNL Current Operations	Proposed BP Solar 37 MW Project
		actions would reduce particulate emissions. *Operation* – None.
Climate	Fossil fuel plant GHG emissions would contribute to potential cumulative effects on climate change.	*Construction* – Temporary increase in vehicle exhaust emissions during construction would minimally increase greenhouse gas (GHG) emissions; removing 153 acres (62 hectares) of trees and vegetation would eliminate absorption/storage source, but minimally impact climate. *Operation* – The 40-year project lifetime avoids emissions of about 1,238,015 metric tons of CO_2 into the atmosphere from conventional generating means; overall beneficial effect on GHG emissions.
Visual Quality	No change from the existing BNL site conditions.	*Operation* – Southern edge of the south array would be visible from passing LIRR trains for approximately 1,100 feet (335 meters) and from residential properties located immediately south of the tracks, (primarily when the trees are bare); array would be visible from aircraft. Overall, the project would not be expected to have an adverse visual impact either on or off-site.
Noise	No change from the existing BNL site conditions.	*Construction* – Temporary increase in noise levels would minimally impact residential properties immediately south of the LIRR tracks. *Operation* – None.
Industrial Safety and Occupational Health	No change from the existing BNL site conditions.	*Construction* - Hazards typical for mid- to large-scale construction activity such as electrical, mechanical, noise and lifting – Hazards minimized by adherence to federal, state, and local regulations, Occupational Safety and Health Administration (OSHA) regulations, general contractor safety plans applicable electrical and fire codes, etc. *Operation* – Common electrical safety hazards - The remote location of the site, the fenced enclosure and warning placards would minimize exposure of the BNL staff, visitors, and public to

Table 1. (Continued)

Comparison Factors	No Action: BNL Current Operations	Proposed BP Solar 37 MW Project
		potential electrical safety hazards. *Use of Brownfields* - Potential exposure to contaminants - Ballasted solar arrays would be used over capped land fill areas. Areas were remediated to industrial standard and proposed use is within industrial use identified in BNL Land Use and Institutional Controls Program - Work planning and control measures would further reduce risk.
Radiological Characteristics	No change from the existing BNL site conditions.	*Construction and Operation* – None. *Use of Brownfields* – Potential exposure to contaminants - Ballasted solar arrays would be used over capped landfill areas; areas were remediated to industrial standards - Work control measures further reduce any potential risk in other brownfield areas As Low As Reasonably Achievable (ALARA).
Natural Hazards	No change from the existing BNL site conditions.	*Operation* – Potential structural failure of an array due to natural phenomenon such as earthquakes, hurricanes, or wildfire could result in physical damage to the array. There is low probability of a hurricane and very low probability of an earthquake; construction is to building code standards; solar arrays do not contain any hazardous materials; located in very low population area; existing wildfire controls are adequate.
Intentional Destructive Acts	No change from the existing BNL site conditions.	The Proposed Action would not offer any targets of opportunity for terrorists; additionally, the potential for vandalism is reduced by fencing, random security patrols and inspections.
Utilities	The LIPA electric grid would not receive 37 MW of electricity produced by solar PV arrays. The BNL internal electric grid would not receive 1 to 2.	*Construction* – None. *Operation* - Electrical power from the solar arrays would be connected to the regional grid via the LIPA substation located adjacent to the Laboratory's south border. The LIPA electric grid would receive up to 37 MW of electricity. Laboratory dedicated array

Comparison Factors	No Action: BNL Current Operations	Proposed BP Solar 37 MW Project
	MW of electricity produced by solar PV arrays	would potentially provide 1-2 MW of solar generated power to the Laboratory's internal electrical grid.
Electric and Magnetic Fields (EMF)	No change from the existing BNL site conditions and operations.	EMF would be well below New York State Public Service Commission guidelines.
Waste Management and Pollution Prevention (P2)	No change from the existing BNL site conditions and operations.	*Construction* - One-time construction wastes including cleared trees and brush, concrete and steel debris from obsolete structural features - Trees suitable for timber would be recycled; unsuitable tree material would be mulched on-site or composted for topsoil; any hazardous materials (such as temporary fuel storages) would be managed under Suffolk County Article 12 controls. *Operation* – None.
Commitment of Resources	No change from the existing BNL site conditions.	*Construction* – Temporary increase in fuel demand for construction machinery. *Operation* - Reduced need on utility grid for electricity generated through conventional means. Subsequent reductions in the demand for fossil fuels, and water resources to support these conventional generating methods; slight increase in water demand, estimated at 500,000 gallons per year (0.1 percent of current BNL water usage), for cleaning arrays.
Decommissioning and Restoration	Not applicable.	Structures and fencing removed and materials recycled; land would be available for future BNL operations or remain open to natural re-growth/ restoration; restoration would utilize native pine barrens species in accordance with BNL Natural Resource Management Plan.

In summary, the analysis of potential environmental effects from the construction and operation of the BP solar array project indicates a net positive benefit to the environment. The key negative environmental impacts to the ecological resources (i.e. trees, endangered species, and migratory birds) are due to the land disturbance. Some of these impacts are minimal and may be reduced with mitigative actions proposed. The positive aspects include promotion of clean energy and future reduction of carbon emissions, improved management of invasive plants, the maintenance of a habitat conducive for endangered and common species, and local socioeconomic advantages during site construction and operation.

2.0. Purpose and Need

The proposed action is for DOE to grant an easement for a solar photovoltaic project by BP Solar, using BNL as a host site. The solar arrays would produce clean, renewable energy and support the DOE's mission to reduce dependence on fossil fuels, facilitate research and development of renewable energy technologies and advance development of energy efficiency. The United States considers energy independence a top national priority and has committed to reducing its need for foreign energy sources and the burning of fossil fuels that add to production of GHGs which result in global climate change.

The proposed project encompasses a large-scale commercial solar PV array totaling approximately 37 MW and potentially a 1 to 2 MW Laboratory dedicated array located on the campus at BNL. The electricity generated from the proposed 37 MW array would be transmitted from the array to LIPA which publicly solicited companies for proposals to provide needed solar-generated power on Long Island. DOE volunteered the BNL federal property as a host site for a proposed project. LIPA's interest in BP Solar's proposal led BP Solar, in turn, to seek DOE approval for access to construct and operate its solar arrays on the BNL federal property.

The proposed solar project at BNL promotes the use of non-polluting, renewable solar energy technology in the largest-scale photovoltaic project in the Northeast. In addition to generating electric power, the project would provide data and practical experience in areas such as PV efficiency, batteries, energy storage, climate, and environmental impacts. This data would be vital to the research and development efforts designed to improve solar technologies and impacts on large utility electric grids. The project would also provide the opportunity to explore new research collaborations and launch educational and community outreach programs on solar energy as a clean and renewable energy source.

3.0. DESCRIPTION OF ALTERNATIVES, INCLUDING THE PROPOSED ACTION

3.1. Proposed Action

The proposed project involves DOE granting an easement to BP Solar to construct a large- scale commercial solar photovoltaic array of approximately 37 MW which would cover approximately 200 acres (80.94 hectares) of the BNL federal site. Electricity generated by these arrays would be connected into the regional utility power grid. In addition, a Laboratory dedicated array of 1 to 2 MW may be constructed and connected to the on-site BNL electric grid. The arrays would utilize, where possible, areas already cleared (agricultural fields, firebreaks, and brownfields), as well as require clearing of an estimated 153 acres (62 hectares) of trees. A building to support the Laboratory dedicated array may also be constructed.

3.1.1. Project Site Location

The proposed project would cover an area in the south central to southeast and east central portions of the BNL property within the Compatible Growth Area as delineated by the Core Pine Barrens and Compatible Growth Areas line (see Figure 3). The proposed area is divided into north and south sections. The north section is composed of approximately 78.5 acres (32 hectares) and is located in the vicinity of the former experimental agricultural fields. The south section is located just north of the line delineating Core Pine Barrens from Compatible Growth areas and mostly east of First Street. The south section is approximately 111 acres (44.92 hectares). Additionally, brownfield[1] areas composed of the Former Landfill, Interim Landfill, Slit Trench, and Glass Holes is available for possible placement of solar panels and/or Laboratory dedicated array facility. This area is immediately west of the southern portion identified above. The brownfield areas cover approximately 18.26 acres (7.84 hectares) and may be used within the controls established in the BNL Land Use Controls Management Plan [BNL 2009a].

While the overall size of the proposed project area is not expected to change appreciably from that depicted in Figure 3, the specific tracts of land may vary. Any minor changes to the specific tracts of land would be within the scope and established criteria set forth in the EA.

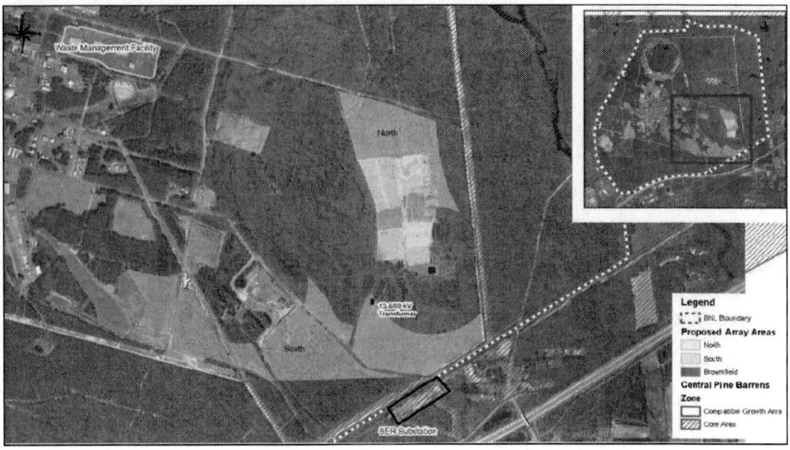

Figure 3. Proposed Areas for BP Solar Project.

3.1.2. BP Solar Arrays

The BP solar arrays would be comprised of individual solar modules, such as the BP3220 modules, or a comparable solar module. The module would have dimensions of approximately 5.5 x 3.3 x 0.016 feet (1,667 x 1,000 x 50mm) weigh 43 lbs (19.4kg), and have a standard rating of 220 watts (W). Approximately 167,712 modules would be used to obtain a total project capability of about 37 MW of direct current (dc). The modules are generally considered to have a lifetime of upwards of 40 years.

The modules would be arranged into sub-arrays consisting of four vertical rows of six modules. The 24 modules would be attached to a steel I-beam support structure that would be anchored with two steel I-beam driven posts. These posts would be driven approximately 15 feet (4.6 meters) below the surface of the soil. The spacing between sub-array rows would be approximately 29 feet (8.84 meters). The gap for shading between the back of one mount and the front of the next would be approximately 18.5 feet (5.64 meters).

Strings of modules would be wired together in series. Fused combiner boxes with disconnect switches would connect strings of modules and lead to 32 concrete equipment pads of about 100 square feet (9.3 square meters) each, placed around the array. Each equipment pad would contain two inverters and one MVA transformer. It is anticipated that the wiring between the modules and equipment pads would be trenched. Conductors would combine at 15kV outdoor, metal-clad, switchgear on a 250 square-foot concrete pad. A 13.8/69 kV transformer, with associated circuit breakers, disconnect switches, and a small (12' x 24') control enclosure mounted on pads totaling 2,000-square-feet,, would then step up the power and supply it via transmission line to the LIPA 8ER substation located across the LIRR track from BNL. The 69kV transmission line would likely be run underground, including three conduit lines, one for fiber optics and two for power distribution (one active and one spare) that would be run under the LIRR tracks. Reference Section 4.1.6 Connected Actions for details on the LIPA Interconnect at substation 8ER. The locations of the 13.8/69 kV transformer and LIPA substation 8ER are shown in Figure 3. A fence designed to comply with the National Electric Code (NEC) requirements would be installed around the perimeter of the solar array areas. Fencing may contain small openings that would allow access for small animals.

3.1.3. Operation and Maintenance

Operation and maintenance of the system would involve replacement of modules, repair of inverters, and other supporting equipment, which would be expected to occur on a limited basis at certain points during the system's estimated 40-year life cycle. The area would be left unplowed from snow during the winter months except for when maintenance is necessary. Modules would likely be cleaned twice the first year. After the first year, modules would be cleaned 0-2 times annually depending on their location relative to pollinating trees and based on the results from the first year. Routine maintenance would also include periodic mowing of the ground cover to discourage invasive species and maintain vegetation heights at less than 2 feet (0.61 meters).

3.1.4. Laboratory Dedicated Array

It is envisioned that a separate array with a generating capacity of approximately 1-2 MW would be constructed by BP for BNL's sole use. This array would serve several purposes but principally would provide BNL with a source of power and allow for research activities related to solar photovoltaic technologies. The array would be tied into BNL's internal electric grid. The research conducted at this dedicated array would be multi-faceted in that BNL anticipates that it would provide climate, grid and efficiency data. It would also potentially serve as a test-bed for energy storage technologies (capacity and cycle testing, and different storage technologies), performance testing of systems in conditions of northern latitudes, module technology comparisons (e.g. multi-crystalline, thin film, etc.), inverter studies (comparisons between manufacturers, technologies and long-term inverter and module testing), shading tests, solar monitor development, ground-mount structural systems development, and possibly tracker testing and development.

The space required for the proposed Laboratory dedicated array is expected to be between 5 and 10 acres and may be developed as one large array or as a segmented array with the potential configuration of 750 kW and 250 kW. Whether the dedicated array is one large array or a segmented array, the BNL dedicated array would be in one location with potentially small spaces between the panels. A support building, of approximately 2,000 square feet (186 square meters), would also be located adjacent to the BNL dedicated array. Potential locations being considered for the BNL dedicated array include the southeast corner of the current Waste Management Facility (WMF) complex, and a portion of the brownfield area described in Section 3.1.1.

The operation of the BNL dedicated array would require the interconnection of the array with the electric grid on-site. This would require the extension of, at a minimum, electric service to either of the two potential sites proposed for this array. Additionally, water and sanitary facilities would be required should a building be necessary for housing bench-top experimental testing and evaluation systems associated with the BNL dedicated array.

3.1.5. Future Upgrades

While no specific upgrades to the solar arrays or project footprint are planned at this time, it would be reasonable to expect that future technological improvements may warrant changing some or all PV panels in order to achieve improved efficiency. It is also conceivable that future upgrades, employing similar updated technologies, could be proposed to expand the arrays beyond their current proposed footprint. Any future upgrades would be reviewed through established NEPA procedure.

3.1.6. Connected Actions

LIPA Interconnect at substation 8ER

The substation (LIPA Substation 8ER) would require very minor modifications to connect the solar energy source to the energy distribution grid. The substation modifications include the addition of two 69 kV disconnect switches, a 69 kV gas circuit breaker, 69 kV potential transformer, revenue metering, metering potential transformers and current transformers, and related control and protection relaying. All modifications would occur entirely within the substation footprint.

The new transmission cable to be installed from the solar arrays located on the BNL site to the LIPA substation will be approximately 900 feet (274 meters) between the BNL property line and the LIPA substation 8ER point of connection. The portion of this cable beneath the LIRR easement (consisting of two conduit lines beneath the LIRR tracks -- one active and one spare) would be constructed below ground using horizontal directional drilling.

Work for Others

The easement agreement between DOE and BP Solar for siting the BP Solar Array project at BNL may include BP Solar providing funding (consideration for the use of the federal property) to the Laboratory to support a Work for Others (WFO) program.

The Work for Others program would allow BSA to perform work for BP Solar, a non-DOE entity, on a fully reimbursable basis. WFO activities may involve use of the Laboratory dedicated array for research described in Section 4.1.4, or other research and development unrelated to solar power. WFO projects are reviewed individually under the existing NEPA review procedure.

Tiger Salamander Breeding Pond Enhancement

The current configuration of the Tiger Salamander breeding pond located, adjacent to the transmission line from the North array field, does not allow it to retain water year-round. The lack of persistent water retention prevents salamanders from developing, thus resulting in a population sink. In order to take advantage of solar array construction resources in the immediate area, the following habitat enhancements would be performed: Existing earthen berms, currently about 3 feet (1 meter) high, and 100 feet (30.5 meters) long, around the pond perimeter would be removed; the area would be regraded to allow water flow into the pond from surrounding wetlands; native vegetation would be planted in the affected area.

3.1.7. Decommissioning and Restoration

Solar PV arrays have an estimated lifetime of about 40 years. At the end of operational life, or substantially reduced efficiency, the arrays would be dismantled and disposed of as part of a planned decommissioning process. All support structures, fencing, and associated electrical hardware (wiring, conduit, towers, etc.) would also be removed. The land would then be available for future BNL operations or restoration, including natural re-growth. The power purchase agreement between BP Solar and LIPA to operate the solar PV arrays at BNL is currently for 20 years.

At the end of 20 years:

- The agreement could be extended for an additional time period.
- The agreement could end as scheduled, whereupon the solar project could then be transferred to another entity to operate or upgrade.
- The agreement could end as scheduled, whereupon the solar facility would be dismantled and the federal property would be restored.

3.2. No Action Alternative

The No Action Alternative would maintain the current conditions and operations on the BNL site.

3.3. Alternatives Considered But Not Assessed

As the DOE and BSA considered the Laboratory site as a potential host location for the photovoltaic systems described in the LIPA Request for Proposal, a number of alternatives were reviewed but rejected. When considering potential locations, DOE and BSA considered three primary criteria for potential sites. These criteria were 1) proximity to the LIPA grid, 2) avoidance of Core Pine Barrens areas, and 3) minimizing impacts to the environment. Additional considerations for placement included minimizing conflicts with future Laboratory development (BNL Master Plan [BNL 2004]) and existing infrastructure. The following alternatives were initially considered and ultimately eliminated for the reasons described below (Reference Figure 4).

Laboratory West Boundary

The West Boundary would include much of the area along the William Floyd Parkway and north of Princeton Avenue. It is within the Compatible Growth Area of the Pine Barrens and does not contain any wetlands. The 377 acre (153 hectare) area is composed of a mix of native pine barrens forests and white pine plantings that protect WWI trenches that are eligible for listing on the National Register of Historic Places (NRHP). This alternative was rejected due to the exposure of the area to traffic on the William Floyd Parkway, disturbance of WW I trench systems, and interference with future Laboratory development. Future Laboratory development in the area potentially includes rerouting of the front gate road, new construction for Facility Operations, and a potential Visitors Facility (BNL Master Plan). The location is also approximately 2.75 miles (4.42 kilometers) from the LIPA substation located on the BNL south boundary along the LIRR.

Laboratory Southern Area

The 585 acre (237 hectare) area south of Princeton Avenue and the south firebreak was considered due to the large expanse of open space with few wetlands. There is limited development, restricted primarily to Upton Road and several access roads for access to groundwater treatment systems. This alternative was rejected because it is wholly within the Core Preservation Area of the Central Pine Barrens and has little fragmentation.

Dispersed Open Areas

There are numerous open areas of lawn, ball fields, and previously disturbed areas associated with past construction or demolition, adding up to approximately 310 acres (125 hectares). Some of these areas include locations within the apartment complex, open fields near the front gate, meteorology fields, and lawn areas around buildings, disturbed spaces within the RHIC ring, the ecology fields, the biology fields, and brownfields. Much of the open space was removed from consideration because it is either currently being developed [National Synchrotron Light Source (NSLS II) – 50 acres (20 hectares)], is scheduled for other construction planned within the BNL Master Plan, may cause significant infrastructure issues (access to or interference with utilities), or may be needed for future uses associated with facility scientific mission [e.g., Relativistic Heavy Ion Collider (RHIC) ring areas] – Reference Table 2 below. Approximately 23.6 acres (9.6 hectares) of the brownfield area was rejected as sites for construction due to issues associated with either residual radiological contamination [15.6 acre (6.3 hectare) former hazardous waste management facility] or configuration of the brownfield that is not conducive to building solar arrays [8 acre (3.2 hectare) current landfill]. Additionally, utilizing all of the dispersed open areas would require development of numerous segregated small solar arrays tied together and then connected to the LIPA substation located approximately 1.75 miles (2.82 kilometers) from the center of the Laboratory's core developed area. Due to the combination of distance, segregation, and existing infrastructure interference, having numerous segregated arrays would not be within the developer's feasible project parameters.

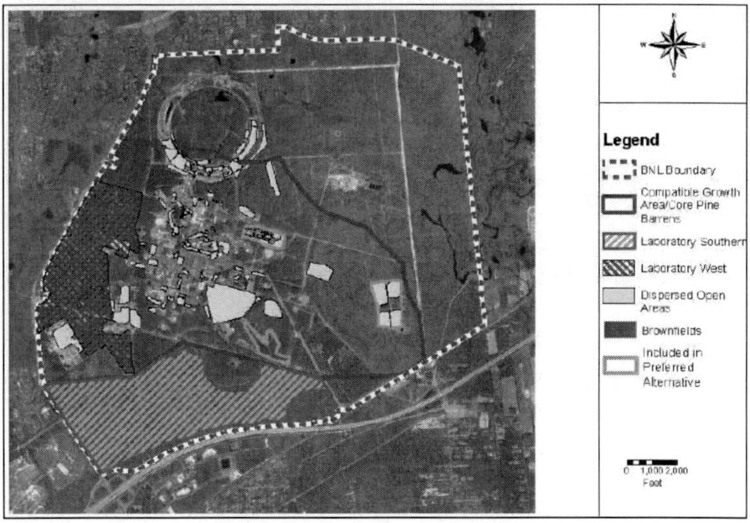

Figure 4. Alternatives Considered But Not Assessed.

Of the 310 acres (125 hectares) of open areas, about 58 acres (23.5 hectares) are included within the Preferred Alternative, encompassing 31.6 acres of Biology Fields, 8 acres associated with the current Waste Management Facility, and 18.3 acres of brownfields. The remaining 252 acres were not assessed further for the reasons mentioned above.

Rooftop Deployment

While BNL has numerous flat rooftops that could be utilized for the installation of solar arrays, this was not considered viable since the proposed BP Solar project is designed to generate 37 MW of electricity for LIPA and not for the Laboratory. This method of deployment would require extensive modifications to roof structures to support the weight of the arrays and use of rooftops would also require that energy generated be directed through existing meters on the buildings (net metering), or that a subgrid system with numerous electric junctures be established to allow generated energy to enter the LIPA grid. A diffuse sub-grid system would create substantial difficulties coexisting with BNL's extensive utility systems. Further, it would require major construction issues trying to install the required infrastructure around BNL's underground utility systems (electric, steam, sanitary, storm, water, communications, etc.)

Table 2. Dispersed Open Areas

Area	Acres (hectares)	Limitation
NSLS-II project site	50 (20 hectares)	Currently being developed
Brownfields	23.6 acres (9.6 hectares)	Residual contamination and/or site contour
Miscellaneous	142.5 acres (58 hectares)	Combination of: Planned/future scientific or support mission; interference with existing infrastructure; distance/ segregation of arrays
Main entrance and apartment area lawns	36 acres (14.6 hectares)	Distance/segregation of arrays
Biology Fields	31.6 acres (12.8 hectares)	Included within Preferred Alternative
Current Waste Management Facility	8 acres (3.2 hectares)	Included within Preferred Alternative
Brownfields	18.3 acres	Included within Preferred Alternative

4.0. Affected Environment and Environmental Impacts

This section describes the general environment in the areas proposed for installation of the photovoltaic arrays, along with specific environmental elements that may be affected by the proposed action. The effects of the preferred alternative on each of these elements are presented within each subsection. Section 5.23 describes the environmental effects of the No Action Alternative. For additional information on the BNL site, including detailed environmental monitoring results, please refer to the Annual Site Environmental Report (SER).

4.1. Site Description

The BNL site encompasses a total of 5,265 acres (2,131 hectares) with most principal facilities located near its central developed area, which occupies approximately 1,656 acres (670 hectares). The balance 3,609 acres

(1,460 hectares) of the site are largely wooded and part of the Long Island Pine Barrens. The central portion of the BNL site is within the compatible growth area as designated by the Central Pine Barrens Joint Planning and Policy Commission (Pine Barrens Commission), while the areas outside the central portions of the Laboratory are designated as Core Preservation Area by the Commission. BNL, as a federal site, is not bound by NY State Environmental Conservation Law Article 57 establishing the Central Pine Barrens. However, DOE works within the spirit of the law whenever possible.

The proposed location for the BP Solar arrays is in the south-central and southeast portion of the BNL site just west of the eastern-most fire-break and just north of the southern-most fire- break (see figure 3) and is contained within the Pine Barrens Compatible Growth Area (see section 5.2). A portion of the area has been used as experimental agricultural and forestry fields, for farming prior, and as an Army camp (Camp Upton) prior to the BNL site being acquired by the Atomic Energy Commission in 1947. In addition, a small observatory building, about 120 square feet (11 square meters) in size, and previously use by the former BNL Astronomy Club, is located in the eastern experimental field. The observatory would be relocated prior to or removed as part of the proposed action. Other parts of the proposed site consist of white pine/fir trees that were planted over the former WWI Remount Facility and areas of native Pine Barrens habitat. The proposed construction site is gently sloped, with most of the area ranging in elevations from 40 to 90 feet (12–18 meters) above sea level. The geologic makeup of the project site (and the Laboratory) is primarily glacial sands.

A potential site being considered for the Laboratory dedicated array is an 8-acre (3.2 hectare) area located in the southeast corner of the current Waste Management Facility (WMF) complex – Reference Figure 3. This fenced area consists of planted/maintained grass and asphalt, and has water, sanitary, and electrical utilities. Another site being considered for the Laboratory dedicated array is the Chemical/Animal Pits and Glass Holes area described in the Use of Brownfields section below.

Use of Brownfields

Additional areas under consideration for siting solar arrays include two brownfield locations (Reference Figure 3.) The Former Landfill Areas is about 10 acres (4.85 hectares) encompassing the Former Landfill, Interim Landfill, and Slit Trench. The Chemical/Animal Pits and Glass Holes Area is approximately 8 acres (3.2 hectares). These areas have undergone environmental restoration under the Comprehensive Environmental Response,

Compensation & Liability Act (CERCLA) and are currently managed by the BNL Land Use and Institutional Controls Program. The Former Landfill Areas were capped with impermeable geomembrane fabric to prevent precipitation from entering the landfill waste and possibly leaching contaminants into the groundwater [BNL 2009b]. Approximately two feet (0.61 meters) of soil covering the cap serves as a protective barrier. Radiological and mercury soil contamination in the Chemical/Animal Pits and Glass Holes Area were remediated to industrial standards.

These brownfield areas primarily consist of open landscape surrounded by woodlands, with the exception of the National Synchrotron Light Source-II (NSLS-II) construction-site located to the northwest. Elevations in the brownfields range from a high of about 85 ft (26 meters) above sea level at the top of the Former Landfill mound, gradually sloping to the southeast to approximately 72 feet (22 meters) above sea level. Grasses 1-3 feet (0.3-1 meters) high cover the landfill mounds, while the other areas are essentially barren or have minimal vegetation.

4.2. Ecology

4.2.1. Existing Environment

The Laboratory has a comprehensive understanding of the various ecological resources present on-site through multiple efforts including an extensive biological investigation conducted in the early to mid-1990s called the Site Wide Biological Survey; the establishment of a Wildlife Management Plan in 1999; the Natural Resource Management Plan in 2003; the establishment of the Upton Ecological & Research Reserve (Upton Reserve) in 2000; and the subsequent studies conducted under both the Upton Reserve and Natural Resources Program as well as volunteer work conducted by the not-for-profit Foundation for Ecological Research in the Northeast (FERN), a non-profit organization. Additionally, from May-July 2009, BNL staff and biologists walked through the area to further examine the fauna and found that conditions had not appreciably changed from prior studies.

Vegetation

Vegetation at BNL is for the most part typical of the pine barrens in which the site is situated. A 2003 aerial photo analysis of vegetation on-site identified 12 vegetation classes. The proposed site for the BP Solar project consists of several different habitats (the corresponding vegetation classifications are included in parenthesis) including: white pine/fir/spruce forest planted by the Civilian Conservation Corps between 1934 and 1941 (planted white pine forest); experimental tree nursery with now fully matured trees consisting of several species of conifers and other trees (planted white pine forest); native pine barrens habitats (a mix of pitch pine/oak/red maple forests); old farm fields undergoing secondary succession (pitch pine/oak/red maple complexes); areas with heavy infestations of invasive plants which exist throughout the area but primarily adjacent to roads and paths; open experimental agricultural fields (grass and successional), and unpaved roads and firebreaks. Because much of the area has been disturbed by either agriculture, WW I and WW II activities, and dense forest plantings, the understory is much less developed than in pine barrens areas with little historic disturbance. The area, including wetlands to the west and north of the proposed project area, provides suitable habitat for numerous protected species, migratory songbirds, raptors, turkeys, reptiles and amphibians, small and medium sized mammals, and deer.

Invasive Species

The area of the proposed project contains several invasive species including Japanese Barberry (*Berberis thunbergii*), Japanese Honeysuckle (*Lonicera japonica*), other honeysuckles (*Lonicera spp.*), black locust (*Robinia pseudoacacia*), phragmites (*Phragmites australis*), black and/or pale swallow-wort (*Cynanchum spp.*), and potentially kudzu (*Pueraria lobata*). These species were either intentionally introduced to the area as ornamentals (i.e. Japanese Barberry), inadvertently transported to Long Island and the BNL site by visitors, or transferred through movement by animals. The area along Brookhaven Avenue and roads within the Biology Fields area are heavily infested with invasives, and most fire-break areas within the proposed project site have Japanese Barberry and black locust present.

Threatened, Endangered, or Species of Concern

There are no known federal threatened or endangered species on BNL property. The NY State designated endangered eastern tiger salamander (*Ambystoma t. tigrinum*) inhabits a constructed wetland and periodically uses several accessory wetlands. Suitable tiger salamander habitat is protected by a 1,000 foot buffer around the primary habitat. Species listed by NY State as species of special concern using the constructed wetland and wetlands to the west and north of the proposed project include the eastern spadefoot toad (*Scaphiophus holbrokii*) and the marbled salamander (*Ambystoma opacum*). Other species of special concern in the area of the proposed project include:

the eastern box turtle (*Terrapene carolina*), possibly the eastern hognosed snake (*Heterodon platyrhinos*), Cooper's hawk (*Accipiter cooperi*), and sharp-shinned hawk (*Accipiter striatus*). Three protected insect species potentially use the area adjacent to the proposed project. They include the frosted elfin butterfly (*Callophrys iris*) which historically used wild lupine (*Lupinus perennis*) along the south fire-break for breeding, mottled dusky wing butterfly (*Erynnis martialis*), a species of special concern, and the Persius dusky wing butterfly (*Erynnis persius*); a NY State designated threatened species. The two dusky wing butterflies have not been documented at BNL but utilize similar plant species as the frosted elfin and were historically in the Pine Barrens. The coastal barrens buckmoth (*Hemileuca maia maia*) may also be found in pine barrens habitat. However, this moth species requires dense stands of shrub oak (*Quercus ilicifolia*), which is not prevalent in the proposed project area. In addition, recent surveys provided no indication of the buck moth within the proposed area. No NYS threatened or endangered plant species are known to exist within the proposed project area.

Migratory Birds

Under the Laboratory's Natural Resource Management Plan, bird surveys have been conducted through all of the major habitat types on site. Surveys have been conducted March through September annually since 2000. Over the nine years of data collection a total of 110 species have been documented. Additionally, birding has been an avid pastime for many BNL employees. Between 1948 and the present, more than 185 bird species have been documented on-site and approximately 85 species routinely utilize the BNL Site for nesting. Of the six bird survey transects located on BNL, two cover habitats similar to those in the area proposed for siting the proposed solar arrays.

The planted pine/fir forests along the southern part of the proposed project site typically have around 27 species with around 56 species using this forest type over the past nine years. The most common bird species in this forest type include black- capped chickadee (*Poecile atricapillus*), American robin (*Turdus migratorius*), American crow (*Corvus brachyrhyncos*), bluejay (*Cyanocitta cristata*), chipping sparrow (*Spizella passerina*) , pine warbler (*Dendroica pinus*), ovenbird (*Seiurus aurocapillus*), eastern or rufous-sided towhee (*Pipilo erythrophthalmus*), eastern wood peewee (*Contopus virens*), tufted titmouse (*Baeolophus bicolor*), and goldfinch (*Carduelis tristis*). Raptors using this forest type include red-tailed hawk (*Buteo jamaicensis*), Cooper's hawk (*Accipiter cooperii*), sharp-shinned hawk (*Accipiter striatus*), and great horned owl (*Bubo virginianus*). The eastern screech owl (*Otus asio*) is also likely to use this forest type. Woodpeckers using this forest type include red-bellied (*Melanerpes carolinus*), downy (*Picoides pubescens*), hairy (*Picoides villosus*), and northern flicker (*Colaptes auratus*).

The mixed habitat of open fields, wet forest, and upland forest associated with the northern area of the proposed project provide suitable habitat for a more diverse species assemblage. Between 2000 and 2008, 95 species of bird have been identified during surveys in this area with an average of approximately 50 species per year seen or heard. Species routinely identified in this area include those listed above and eastern bluebird (*Sialia sialis*), field sparrow (*Spizella pusilla*), great crested flycatcher (*Myarchus crinatus*), indigo bunting (*Passerina cyanea*), house wren (*Troglodytes aedon*), northern cardinal (*Cardinalis cardinalis*), red-eyed vireo (*Vireo olivaceus*), red-winged blackbird (*Agelaius phoeniceus*), scarlet tanager (*Piranga olivacea*), tree swallow (*Tachycineta bicolor*), wood thrush (*Hylocichla mustelina*), and yellow-billed cuckoo (*Coccyzus americanus*). In addition to the raptors mentioned above, American kestrel (*Falco sparverius*), merlins (*Falco columbarius*), and eastern screech owl (*Otus asio*) have been seen using the area.

Mammals

A number of mammals utilize the various habitats at BNL, including those areas proposed for solar arrays. The largest mammal found at BNL is the white-tailed deer (*Odocoileus virginianus*), which is present in numbers exceeding 50 per square mile (19.31 per sq. kilometer). In general, there are fewer deer within the planted pine/fir plantations compared with the areas around the experimental agricultural fields, which have some of the densest deer population on-site. Other mammals on-site include small mammals (bats, mice, squirrels, rabbits) and medium-sized mammals like raccoons and red and grey fox.

Only two bat species have been confirmed on-site, the little brown bat (*Myotis lucifugus*) and the eastern red bat (*Lasiurus borealis*). The little brown bat typically utilizes buildings for maternity colonies and utilizes trees along streams and water bodies for summer foraging. This bat has been seen utilizing open water areas on- site. The eastern red bat utilizes trees and forested areas for roosting and foraging. This bat is occasionally seen on-site.

Small mammal species found on-site include the meadow jumping mouse (*Zapus hudsonius*), white-footed mouse (*Peromyscus leucopus*), least shrew (*Cryptotis parva*), short-tailed shrew (*Blarina brevicauda*), meadow vole

(*Microtus pennsylvanicus*), house mouse (*Mus musculus*), eastern chipmunk (*Tamias striatus*), grey squirrel (*Sciurus carolinensis*), southern flying squirrel (*Glaucomys volans*), groundhog (*Marmota monax*) and eastern cottontail rabbit (*Sylvilagus floridanus*).

Reptiles and Amphibians

The BNL site is home to 28 species of reptiles and amphibians. The various species are distributed throughout the Laboratory site but may be localized depending on their habitat requirements. Reptiles like the eastern box turtle (*Terrapene carolinensis*) may be found in virtually all habitats on-site, while many species of snakes and other turtles are localized near wetland resources. Frogs and toads are isolated around wetlands during breeding periods but may be found moving away from wetlands to forage for food during the late spring through summer months. Several salamander species can be found in and adjacent to wetland areas on-site. These salamanders include the NY State designated endangered eastern tiger salamander (*Ambystoma t. tigrinum*), marbled salamander (*A. opacum*), red-spotted newt (*Notophthalmus viridescens*), and red-backed salamander (*Plethodon cinereus*).

Pine Barrens

The BNL site is within the Central Pine Barrens of Long Island. This area has been designated a protected area under NY State Environmental Conservation Law Article 57. While federal lands within the Central Pine Barrens are not bound by the law, the DOE and the Laboratory attempt to work within the spirit of the law when planning and implementing projects. The Central Pine Barrens is an area of approximately 105,000 acres and is divided into a Core Preservation Area of approximately 55,000 acres where development is proscribed and limited, and the Compatible Growth Area (CGA) of approximately 50,000 acres, where development is allowed but must meet a series of standards and guidelines established in the Land Use Plan for the Central Pine Barrens [CPB]. The proposed photovoltaic project at BNL is entirely within the CGA and is expected to meet the established standards and guidelines.

4.2.2. Effects of Preferred Alternative on Ecological Resources

Effects on Vegetation during Construction

The proposed project would result in disturbance to approximately 200 acres (80.94 hectares) of land through clearing and grading in preparation for construction. The construction would follow a phased approach in which enough area to construct 3 megawatts of generating capacity would be completed at a time (approximately 15 acres or 6.07 hectares). As the construction of one phase is started, the clearing for the next phase would begin. Overall, construction would impact vegetation within the proposed areas with the majority of effects on trees (overstory).

Based on existing construction design, trees would be cut, stumps removed and holes filled in, with as little disturbance to understory vegetation as possible. The initial effects would be removal of trees. The age of the oldest trees within the proposed project area is estimated to be about 80 years. An estimated 153 acres (62 hectares) of trees would be removed, which constitutes approximately 3% of the entire BNL site and approximately 4% of the Laboratory's forested area. More specifically, the clearing for construction would result in 21.5 acres (8.7 hectares) of Civilian Conservation Corps (CCC) planting being removed compared to 468 acres (189.4 hectares) existing on the BNL site; 129.5 acres (52 hectares) of pine barrens forests compared to 3,532 acres (1,429 hectares) of this forest type on the BNL site; and the removal of 4 acres (1.62 hectares) of matured nursery trees. Removing the overstory would be considered a permanent impact during the estimated 40 year project lifetime as the area of the solar arrays would have to be maintained free of trees to prevent shading of the arrays. Elimination of trees, overstory layer, within the array areas displaces tree nesting birds and tree dependent small mammals to other forested areas on the BNL property.

Clearing would likely have a temporary impact on understory plants, but most would recover as long as the root mass is not disturbed and soils are not heavily compacted due to movement of construction equipment. Under existing construction plans, areas where grading is required to meet specifications, top soils would be removed, grading accomplished and top soils replaced with subsequent planting of native grasses or understory vegetation, such as huckleberry and low-bush blueberry. Areas that currently lack understory vegetation due to dense forest canopy would have native grasses, lupine, or understory vegetation planted.

Additionally, clearing would include the removal of various invasive plant species that if not removed could result in operational and maintenance problems in the future. The removal of invasive plants from the proposed area decreases the potential that these plants would expand further into the Pine Barrens to the east of the proposed area or into adjacent wetland areas.

Effects on Vegetation during Operation

Since operation of the proposed solar arrays is dependent on an open view of the sky, vegetation management would be an integral part of maintenance activities within the proposed project area. Various tree species would likely grow from root material left behind during the clearing process. These "volunteer" trees would be periodically removed. Various understory plants may grow taller than the estimated two-foot height limit for vegetation under the solar arrays. When plants approach that height limit, mowing or trimming would be required. Grasses would likely be cut back to a few inches high, while understory plants like low bush blueberry and black huckleberry would be cut back to heights of 6 – 12 inches (0.15 – 0.3 meters) high. Mowing would not be expected to occur more than twice per year for grasses and once per year for understory vegetation. Increase in density and variety of understory plants would improve this vegetation type which provides habitat for a host of other species. The proposed project would likely have long-term positive effects for these plant species.

While most invasive plants in the area would be removed during clearing, many may persist through remnants of roots or seeds within soils and would have to be managed to minimize effects on the solar arrays and the native vegetation used under the arrays. Continued long-term management of invasive plants would provide assurance of control that would prevent expansion of invasives to natural habitats in the areas surrounding the arrays. Control of invasives may be by manual, mechanical, or chemical means. Use of herbicides may require special permits from NYSDEC in areas in proximity to wetlands. Laboratory policy is to utilize Integrated Pest Management (IPM) as an approach in minimizing pesticides use. Use of IPM would result in long-term reductions in the number and type of invasive species that could move into other pine barrens habitats.

Effects on Threatened, Endangered, or Species of Concern during Construction

A portion of the South end of the North Array, about 12 acres (5 hectares), would be constructed within the 1,000 foot (304.8 meters) protection zone surrounding a constructed wetland designed as eastern tiger salamander habitat. Because this area currently consists of open fields, with additional fields to the north, it is not considered suitable for tiger salamanders. Approximately four acres (2 hectares) of solar panels in the South Array would also be located slightly within the 1,000 foot (304.8 meters) protection zone of a known tiger salamander habitat. A barren utility right of way (Power

Line Road) situated between the proposed project area and salamander habitat make the site unsuitable for tiger salamanders. NYS draft guidance for eastern tiger salamander breeding ponds recommends that 50 percent of the area within 1,000 feet (304.8 meters) and 100 percent of the area within 535 feet (163 meters) be maintained as available habitat to the salamanders [NYSDEC].

While arrays would be located within the 1000 foot protection zone of two separate habitat areas, the project would continue to be meeting the NYSDEC protocol. Locating arrays within the two identified protection zones was evaluated and determined to be permissible by the NYSDEC. Recent research indicates these salamanders moving as much as 1,250 feet (381 meters) from breeding ponds to forage (*V. Titus pers. comm.*). Therefore, construction of the proposed solar arrays may have a minor impact on the tiger salamander beyond the 1,000-foot (304.8 meters) protection zones around known habitats primarily consisting of loss of existing wooded habitat, but also potential inadvertent injury/kills. Existing processes, such as digging permits, would identify and inform project personnel of the potential for salamander sightings and instruct them to contact BNL natural resource personnel. Following construction, potential salamander habitat would be reestablished through native vegetation, such as grasses, low bush blueberry, and huckleberry, planted under the solar array canopy. Routing and installation of a transmission cables from the north array to the 13.8/69 kV transformer pad, and then on to the LIPA substation would utilize existing unpaved fire break roads or be performed in areas disturbed for solar array placement (e.g., installation would include trenching and burying the cables under the solar arrays). This alternative was proposed to take advantage of an already disturbed area with no vegetation by trenching down the middle of the existing road and burying the cable. This approach would overlap with a 100 foot (31 meter) wetland buffer area, but avoids an alternative path that would require clearing trees within a 535 foot (163 meter) tiger salamander habitat buffer. Installation actions would require a NYSDEC wetlands permit and would incorporate protective measures such as silt barriers, hay bales, etc.

Following cable installation, the roadway would be returned to its pre-project state, with no impacts anticipated during system operation. Due to the short duration of the trenching and cable installation activity (less than 1 week) over an existing roadway, adverse impacts to the tiger salamander would not be anticipated. Overall, due to established buffers and procedures, construction should result in minimal impacts to tiger salamanders and other amphibians. Disturbance from construction would temporarily impact eastern box turtles.

However, proposed fencing design is expected to allow movement of reptiles throughout the entire area.

Current habitats for threatened and endangered butterfly species are of marginal quality and require disturbance for maintenance. Disturbance from construction may result in potential expansion of appropriate host plants (wild lupine) in areas adjacent to existing plants.

Removal of approximately 153 acres (62 hectares) of mature forest would impact nesting raptors, forcing them to move from the proposed construction-site to other suitable habitat either on BNL property or within nearby pine barrens forest. Cooper's Hawks and other small raptors would likely utilize extensive white pine forests on the BNL site, as would owls. Overall, construction would have a minor negative impact on threatened, endangered, and special concern species.

Effects of Operations on Threatened, Endangered, and Special Concern Species

Operations would have little additional effects and may have slight positive effects on threatened, endangered, and special concern species beyond the effects of construction. Management of native vegetation (mowing, cutting, invasive species removal, etc.) may result in increased numbers of host plants for rare butterflies. Fencing would prevent entry of deer allowing understory vegetation to mature and seed, which may provide improved habitat for many threatened and endangered species.

Effects of Construction on Migratory Birds

Clearing of trees and temporary disturbance of understory vegetation would result in a moderate negative effect on migratory song-birds and raptors. Noise from construction and movement of vehicles and workers may cause disturbance of some nesting birds. Removal of trees would result in some destruction of nests if clearing occurs during late spring and summer months. Timing the clearing of forests can minimize direct impacts. While clearing would remove available habitat, it affects 4.35 percent of the available habitat on the BNL site. Removal of existing fences and open areas for the construction of the proposed solar arrays would have temporary effects on eastern blue birds which utilize nest boxes within the area. Nest boxes would be replaced once construction is completed.

Effects of Operation on Migratory Birds

Operations of the proposed project would likely have slight positive effects on migratory birds due to increased edge (forest to clearing interface) habitat, improved deer free understory, and increased fence lines for placement of nest boxes. Increased edge habitat may result in slight increases in the number of brown-headed cowbirds parasitizing songbird nests.

Effects of Construction on Mammals

Since over-abundant deer populations have already impacted small mammal populations, removal of trees would have minor effects on mammals. Species most likely affected include flying squirrels, grey squirrel, and white-tailed deer that would be dispersed to surrounding forests. Displaced deer would cause added stress on surrounding forests, and would suffer increased stress due to the effects of overabundance. Other medium sized animals would be displaced with little effect on their populations.

Effects of Operations on Mammals

The operation of the proposed project would likely have minor positive effects on small mammals and medium-sized mammals due to fences keeping deer out of the area. As the understory will be improved, food for small mammals will increase and they, in turn, will become food for medium-sized mammals.

Effects of Construction on Reptiles and Amphibians

Construction activities associated with the proposed project are expected to have minimal impacts on reptiles and amphibians. Species most affected may be the Eastern box turtle and the grey tree frog. Box turtles may have difficulty avoiding construction equipment and trees removed would result in loss of forage and mating habitat for tree frogs. Most other species would only be slightly affected since disturbance of understory vegetation would be minimized as much as possible.

Effects of Operations on Reptiles and Amphibians

The operation of the proposed project is expected to result in slight habitat improvements for reptile and amphibian species. Increased habitat for mole salamanders, tree frogs, snakes, and maintained habitat for turtles should result from increased understory. Proposed fencing may be designed to support movement of small animals, including reptiles and amphibians between the proposed project and surrounding habitats.

Effects of the Construction on the Pine Barrens

The clearing of approximately 153 acres (62 hectares) of trees would have a minor impact on the overall quality of the Central Pine Barrens ecosystem. The proposed project is fully within the Compatible Growth Area (CGA) on the BNL site and falls well within the 65% clearing standard established under the Central Pine Barrens Land Use Plan. The proposed solar project increases the cleared area of the BNL site from 26.8% to 29.8%. When the development standards are applied to just the compatible growth area, the increase in clearing changes the cleared areas from 40% to 47% of the CGA on the BNL property which is still well within the 65% clearing standard. The proposed project, if approved, would require various environmental permits including those for: New York State Wild, Scenic, and Recreational Rivers Act (WSRRA), freshwater wetlands, and construction storm water permits under the NY SPDES program. These permits would meet the requirements of the Central Pine Barrens Land Use Plan [CPB].

Effects of Operations on the Pine Barrens

The operation of the proposed project would have little, if any, overall effect on the surrounding Pine Barrens. Since the proposed project intends to utilize native vegetation underneath the solar arrays and would have minimal impermeable (concrete or paved) surfaces, the recharge of precipitation to groundwater would not be affected.

Few, if any, studies have been performed to evaluate the long-term effects of utility- scale solar array complexes on the ecosystem. Presented with this unique opportunity, the Laboratory would initiate long-term monitoring to assess resulting impacts. Potential areas of study would include: Vegetation growth under solar panels (reflective sunlight, water impingement, etc.); effects on small mammals, amphibians, and migratory birds.

4.3. Water

4.3.1. Existing Environment

Water resources associated with BNL include both surface waters and groundwater.

Surface Water

The BNL site lies within the headwaters region of the Peconic River watershed. Liquid effluents from the BNL Sewage Treatment Plant discharging into the Peconic River receive tertiary treatment and conform to the criteria in the approved SPDES permit issued by the NYSDEC.

Pocket seasonal wetlands are also found throughout the site. A constructed wetland designed as habitat for the eastern tiger salamander (*Ambystoma t. tigrinum*) is within the vicinity of the area proposed for this project. Three federal jurisdictional wetlands are located approximately 1,000 feet (308 meters) northwest and up gradient from the proposed project site, and another jurisdictional wetland is just to the west of the project in the east central portion of the BNL property.

Groundwater

The BNL site is situated over a U.S. Environmental Protection Agency (EPA)- designated sole-source aquifer that is the primary source of drinking water for both on- and off-site private and public supply wells, and water for industrial use such as cooling and steam generation. Across the proposed PV array construction area, the ground surface ranges from 40 to 90 feet above mean sea level (AMSL), and the top of the water table ranges from 36 to 40 feet AMSL. Table 3, below, summarizes the land surface and water table elevations for the proposed general solar array areas, with the south array further divided into southeast and southwest sections.

Past spills and waste-handling practices resulted in contamination of the groundwater with volatile organic compounds (VOCs) and radionuclides at various locations on the Laboratory site. The groundwater below a small portion of the proposed southwest PV array area is currently contaminated with low levels of VOCs and strontium-90 that were released to the environment from the nearby Former Hazardous Waste Management Facility and Current Landfill (CERCLA Operable Unit I) and Chemical Holes/Former Land Fill (CERCLA Operable Unit I/IV). This VOC plume is present in the deep portion of the Upper Glacial aquifer, located more than 100 feet below the water table. The strontium-90 contaminated groundwater is generally positioned within the uppermost 20 feet (6 meters) of the aquifer or 14–34 feet (4-10 meters) below land surface [BNL 2008]. The Sr-90 and VOC concentrations are below the 8 pCi/L and 5 ug/L drinking water standards, respectively. The contaminants in this area are undergoing monitored natural attenuation. In the past, the shallow groundwater in several sections of the proposed northern PV array area was contaminated with ethylene dibromide (EDB), which had been used as a fumigant in the Biology Department's agricultural fields (CERCLA Operable Unit VI). Currently, only trace levels of EDB are occasionally detected in this area.

Sixteen groundwater monitoring wells and a former irrigation supply well are located within the footprint of the proposed project area. The monitoring wells are used to periodically sample groundwater contamination plumes.

Peconic River Scenic Corridor

Approximately 40 acres (16 hectares) of the proposed project would be located within the Peconic River Scenic Corridor, which constitutes the area located within a half-mile of the river as designated by the New York State Wild, Scenic, and Recreational Rivers Act under NY Article 15, Title 27, 15-2707. Vegetative cover in the 40 acres consists of the following estimates: 5.5 acres (2.2 hectares) of mature former tree nursery stock, consisting primarily of conifers; 10 acres (6.5 hectares) of former BNL experimental fields, primarily open field grass and successional plants; 14.1 acres (5.7 hectares) of former farm fields undergoing secondary succession, primarily a mix of pitch pine, oak, and red maple; and 10.4 acres (5 hectares) of successional oak-pine mix.

4.3.2. Effects of Preferred Alternative on Water Resources

The proposed BP Solar project is not expected to impact any groundwater or surface water resources, including wetlands. The boundaries of the project are expected to be outside the 100-foot (30.5 meter) wetland buffer established by NYSDEC; with the exception of a transmission cable installed along an existing unpaved fire break road – reference section 5.2.2 *Effects on Threatened, Endangered, or Species of Concern during Construction* for additional details. Construction stormwater controls would be in place to protect wetlands. Overall rainwater infiltration and groundwater flow conditions would not be affected during construction or operations. Due to the depth of the groundwater contamination plumes, construction and operation of solar arrays in these areas would have no effect on personnel health/safety. The former irrigation supply well located in the proposed project area may be abandoned in place in accordance with an established process that conforms to NYSDEC requirements, or it may continue to be used as a source of water for periodically washing the PV arrays.

A New York State Long Island Well Permit would be obtained for well point utilization as appropriate. Groundwater monitoring wells would be evaluated on an individual basis. If a well interferes with installation of a solar array, the well could be abandoned in place, with the Laboratory maintaining the option to install a new well in a more suitable location. If the height of a well casing is an issue, casings extending up to three feet (1 meter) above the ground surface would be modified to 'flush mount' casings extending only 0.5 feet (0.15 meters) high. A sufficient number of wells would remain intact and accessible to sampling and maintenance personnel to ensure adequate groundwater monitoring capabilities. Previously abandoned wells may also be present below grade in the proposed project area. If structural features associated with the abandoned wells were encountered during construction activities, their removal would have no adverse environmental impacts.

During construction, storm water and silt runoff from project areas would be managed in accordance with the pollution prevention plan prepared under a NYS General Storm Water Permit for Construction Activities. Examples of pollution prevention measures include the use of standard erosion control mechanisms such as silt fencing and stabilized rock construction entrances. In addition, the Laboratory's Environmental Protection Division (EPD) staff would perform periodic inspections of the construction site to verify the adequacy of contractor-implemented controls. After installation, native vegetation in the form of a low-growing ground cover would be planted under/around the solar arrays to minimize the potential for soil erosion.

Existing drainage ditches from past agricultural and/or mosquito control currently aid in managing storm water flows from the area west of the constructed wetland. The surface contour and the potential tie-in of the proposed action with these ditches would be reviewed to determine the need for other storm water management methods, such as installation of culverts, water control structures (e.g., gated weir), and open channel flow measuring devices (e.g., Parshall Flume) for estimating flows. This review would also evaluate the need for establishing a new storm water outfall under the existing or new SPDES permit.

Table 3. Depth to Groundwater

Array Area	Land Surface Elevation (AMSL)	Water Table Elevation (AMSL) (1)	Depth to Water Table
Northern	40-60 feet (12-18 meters)	37-40 feet (11-12 meters)	3-20 feet (1-6 meters)
Southeast	50-60 feet (15-18 meters)	36-38 feet (11-12 meters)	14-22 feet (4-7 meters)
Southwest	60-90 feet (18-27 meters)	37-39 feet (11-12 meters)	23-51 feet (7-15 feet)

Note: Water table position based upon November 2008 measurements.

Construction of the following support features would result in a slight increase in impermeable surfaces – a total of approximately 12,890 square feet

(1,178 square meters): some 32 concrete equipment pads, each about 270 square foot (25 square meters) supporting solar array inverters and transformers; a 250 square foot (76 square meter) equipment pad supporting a switchgear equipment building; a 2,000 square feet (232 square meters) concrete equipment pad for the 33MVA step-up transformer; and a 2,000 square foot (186 square meter) Laboratory dedicated array Support Building. Because these are discontinuous impervious surfaces, no adverse impacts to storm water or water infiltration would be anticipated.

Construction and operation of the below-ground electrical conduit lines, approximately 8-10 feet deep (1.2–1.8 meters), connecting the 13.8 kV transmission line from the on-site solar arrays to the existing off-site LIPA substation, would not infiltrate groundwater, which is at depth of 14-22 feet (4-7 meters). Additionally, water for horizontal directional drilling mud would be obtained from existing BNL water supplies or trucked-in sources.

It is estimated that as much as 500,000 gallons of water per year would be required for cleaning PV panels. Because this represents an increase of about 0.1 percent of current BNL site water usage, existing sources would be considered adequate. Water could be supplied from several sources including a fire hydrant located near the Former Waste Management Facility parking area or an existing agricultural well located in the north Biology Fields. Back-flow prevention hardware would be required when drawing water from the system and washing operations would be designed with water conservation efforts in mind.

Peconic River Scenic Corridor

The proposed action would remove about 40 acres (16 hectares) of trees within the Peconic River Scenic Corridor; however, at its closest point the project would remain approximately 2,000 feet (610 meters) from the river. The Wild, Scenic, and Recreational Rivers Act describes scenic rivers as: 'Those rivers, or sections of rivers, that are free of … dispersed human activities which do not substantially interfere with public use and enjoyment of the rivers and their shores'; and stipulates 'management of scenic river areas shall be directed at preserving and restoring the natural scenic qualities of such rivers'. Due to the density of trees separating the river and the project area, the solar panels would not be seen from the river or its shoreline. As a result, the project would have minimal impact on corridor and no impact on the scenic attributes of the river or its shores.

Permits

Because a portion of the proposed northern PV array area is within a half-mile of the Peconic River, and a transmission line is within 100 feet (30.5 meters) of wetlands, a Wild, Scenic and Recreational Rivers Act Permit and a Wetlands Permit would be required. Both permits are issued by the NYSDEC. Additionally, since storm water from the construction site may flow into wetlands to the west of the northern part of the project site, a NYS General Storm Water Permit for Construction Activities may be necessary. All site runoff would be managed in accordance with the pollution prevention plan prepared under that permit. Depending on final construction configuration, a storm water discharge permit would be necessary under NYSDEC SPDES requirements.

Use of Brownfields

Placing PV arrays in brownfield areas that have been capped would require the use of ballasted arrays to ensure that the cap is not penetrated. If placed on brownfields, concrete ballasts would be used to secure the PV arrays in place, eliminating the need to install in-ground anchors. Additional means of assuring cap protection may include restrictions/limitations on soil disturbance, adding soil layers, monitoring of construction techniques, and periodic post-installation inspections.

The ballasted solar arrays would add some noncontiguous impervious surfaces (concrete bases) to certain project areas. Because the ballasts would not result in a continuous impervious surface, water infiltration rates would not be affected and storm water runoff would not require engineered control mechanisms beyond those that currently exist. If the Laboratory dedicated array were located in the Chemical/Animal Pits and Glass Holes Area, additional impervious surfaces would include about 10,890 square feet (3,716 square meters) of parking area/roadway. Storm water management issues would be evaluated through established Laboratory programs.

4.4. Land Use, Demography, and Environmental Justice

4.4.1. Existing Environment

Land Use

Land use to the east within one mile (1.6 kilometers) of the Laboratory consists of preserved open space, public and private land dedicated to public recreation, and low-density residential areas, one dwelling or less per acre. To the north is a mixture of residential properties, commercial retail and service properties, and public utility services. Schools and churches, open space, and low-to-medium density residential areas are found to the west. To the south are commercial and industrial properties, vacant land, and medium-to-high density residential areas of two or more dwellings per acre. On-site land use consists of open space, scientific, industrial and commercial, agricultural, and residential areas. The brownfield areas are designated for industrial use within established controls.

The current BNL site was established in 1947 specifically to develop and construct large-scale scientific facilities. Figure 5 "Land Use Within 1-mile of BNL Border" presents a 2007 aerial photograph of the Laboratory site and surrounding areas.

Demography

Based on the 2000 U.S. Census and associated population estimates for 2007, approximately 13,460 persons live within 1.0 miles (1.6 kilometers) of the Laboratory's boundary [Davis]. Figure 6 "Population Within1 Mile of BNL Border" shows the BNL site boundary and 1-mile extent superimposed over a map of the U.S. Census blocks, along with the estimated populations in 2007.

The Laboratory's on-site population includes approximately 3,000 employees and more than 4,300 guest researchers who visit periodically each year[2]. An average of 180 people live in temporary on-site housing, and an average of 130 guest scientists and students who visit the Laboratory stay in the dormitories.

Environmental Justice

Environmental Justice is the fair treatment and meaningful involvement of all people regardless of race, color, national origin, or income with respect to the development, implementation and enforcement of environmental laws, regulations and policies.

Fair treatment means that no group of people, including a racial, ethnic, or socioeconomic group, should bear a disproportionate share of the adverse environmental consequences resulting from industrial, municipal, and commercial operations or the execution of federal, state, local and tribal programs and policies.

Federal agencies must identify and address disproportionately high and adverse effects of federal projects on the health or environment on minority and low-income populations (Executive Order 12898). An environmental justice population is defined as a population being at least half minority status or at least half low-income status, or this status is meaningfully greater than the general population.

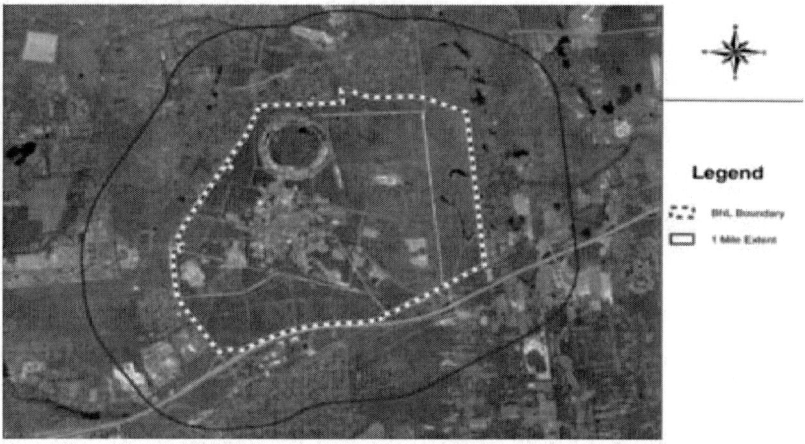

Figure 5. Land Use Within 1 Mile of BNL Border.

Figure 6. Population Within 1 Mile of BNL Border.

Table 4. Low Income Status in Communities Adjacent to BNL Site

Town or Community	Population (2000)	Poverty Status in 1999 – Families*	Population in Poverty Status
Brookhaven Town	448,248	3.9 %	17,482
Ridge	13,380	4.4 %	589
Shirley	25,395	5.5 %	1,397
Manorville	11,131	2.4 %	267
Yaphank	5,025	3.3 %	166
Combined total (Communities Only)	54,931	4.4 %	2,419

* The U.S. Census Bureau defined the average poverty threshold as a maximum annual income of $16,895 or less for a family of four for the year 1999 [U.S. Census, 1999].

A minority is defined as Black or African-American, Hispanic or Latino, Asian, American Indian and Alaskan Native, Native Hawaiian and other Pacific Islander. BNL is situated within the Town of Brookhaven which has a population of 489,255 persons, based on the 2007 LIPA Long Island Population Survey [LIPA], (448,248 based on 2000 U.S. Census data). According to the 2000 U.S. Census data, 15.2 percent of Brookhaven Town's population consisted of minorities [SC]. Using the same 2000 U.S. Census data [Tele Atlas], within 1-mile of the Laboratory's boundary the percentage of minority population is estimated to be approximately 15.9 percent [Davis]. While the percentage of minorities is slightly higher than that of the Brookhaven Town, the 0.7 percent difference would not constitute a percentage that is meaningfully greater than the general population. Therefore, the population living within 1-mile of the Laboratory border would not be defined as an environmental justice population based on minority status.

In regard to low-income status, no data was available to evaluate the income level of the discrete population living within 1-mile of the Laboratory's boundary, or corresponding to the same geographic blocks used for the population data. Income data for the year 1999 was available for specific geographic communities adjacent to the BNL boundary through the Suffolk County government website [SC]. It must be noted that these communities extend six to eight miles beyond the site boundary, and encompass a much larger population than the areas associated with the population data. In two of the four communities evaluated, the percentage of low-income families was slightly higher than that of Brookhaven Town.

Evaluating the combined population of the four adjacent communities, the poverty status for families is approximately 4.4 percent.

While the percentage of low-income families is slightly higher for the combined populations of the four communities bordering BNL than that of Brookhaven Town, the 0.5 percent difference may not constitute a percentage that is meaningfully greater than the general population. Therefore, the population living within 1-mile of the Laboratory border would not be defined as an environmental justice population based on low-income status.

4.4.2. Effects of Preferred Alternative on Land Use and Demography

Land use within the proposed project footprint would change from currently forested and open areas to groups of fenced solar arrays. The effects of installation and operation of the proposed solar arrays would not change the type of activities/operations performed at other areas of the BNL site, and would have no effect on off-site land use or demographics. Operation of the Laboratory dedicated array would result in a small increase in scientific and support staff estimated at less than 20 individuals, along with visiting research personnel.

Use of Brownfields

Siting solar arrays on the brownfield locations identified in Section 5.1 would be within established industrial uses identified in the BNL Land Use and Institutional Controls Program. Utilizing the brownfields would not alter the land use activities/operations performed at other areas of the BNL site or the demography on- site or off-site. Utilizing the open brownfield areas would valuably utilize formerly- contaminated land and further provide an ecological benefit by reducing the potential amount of forested land needed for the proposed solar arrays, while continuing to ensure their long-term protection.

Environmental Justice

The analysis indicates that the proposed action would not be located in the vicinity of a population having a meaningfully higher percentage of minorities or low-income persons. Additionally, the Preferred Alternative, including potential use of brownfield areas, would have no environmental justice impacts because there would be no anticipated negative economic or health effects on any potentially affected population. Therefore, there would be no disproportionate impacts to either low- income or minority populations.

4.5. Socioeconomic Factors

Socioeconomic factors describe the local economy and employment that may be influenced by the Proposed Action.

4.5.1. Existing Environment

BSA employs approximately 3,000 full and part-time personnel and has over 4,300 visiting scientific researchers annually. An additional 40,000 members of the public visit the Laboratory site each year as part of educational and group tours, conferences and events. Direct spending of $454.4 million by BSA in fiscal year 2004 caused a total output of goods and services to the region to expand by more than $800 million. It is estimated that earnings increased by more than $308 million and more than 7,700 secondary jobs were created throughout the economy. Projected spending for fiscal years 2005-2014 could total almost $5.6 billion. More than 91,000 jobs would be created, and virtually all industries, including some of the state's key manufacturing industries, would benefit from spending by the Laboratory [Kamer].

4.5.2. Effects of Preferred Alternative on Socioeconomic Factors

Total project funding of the proposed action, estimated at $200-250 million, would directly impact the local and regional economies. Secondary economic benefits would also be realized through increased personal spending, wages and the spending of non-local workers during their stay in the area.

It is estimated that the proposed action would require a workforce of approximately 200 full-time employees during the construction phase, with two full-time operations and maintenance managers after system start up.

4.6. Transportation Conditions

4.6.1. Existing Environment

Laboratory staff and the majority of visitors commute in their own private or rental vehicles. The Laboratory operates and maintains a fleet of approximately 340 vehicles, ranging from cars and light trucks to delivery, construction, and heavy equipment machines. Included in the BNL fleet are 77 alternative-fuel vehicles, which account for 48% of the light duty vehicles and roughly 23% of all of the vehicles. The general public is restricted from access to the BNL site unless participating in a scheduled event. Commercial delivery, construction and service contractor vehicles are permitted access to the site as necessary.

Routine access to the BNL site from surrounding areas is available primarily through several major roadways including the Long Island Expressway (LIE or I-495) and the four-lane divided William Floyd Parkway (County Road 46). Normal entry/egress is through the Main Gate located at the intersection of Longwood Road and William Floyd Parkway, along the western border of the BNL site. The BNL site can also be accessed through a normally locked gate located on the fire-break road in the southeast corner of the property. The areas proposed for siting the solar arrays are easily accessible through existing unpaved firebreak and access roads. The potentially affected areas also encompass a network of these fire-break and access roads.

The Laboratory maintains an on-site railroad spur branching off the LIRR. The spur is primarily used for transporting waste off-site, but could also be utilized for material/equipment delivery.

4.6.2. Effects of Preferred Alternative on Transportation Conditions

Preferred Alternative

Construction activities associated with the preferred alternative would result in a temporary increase in the number of vehicles entering and exiting the BNL site each day, including workers, material deliveries, and waste transport. The magnitude of the increase would vary depending on the timing of the specific construction phase. During clearing operations, heavy construction vehicles such as bulldozers and tree- harvesting machines would be brought on-site and remain only for their specific period of use.

Preparatory and construction activities may be conducted simultaneously in separate sections. For example, clearing and land preparation would be initiated on a specific section. Once that section is cleared and prepped, array construction would commence and the clearing/prep work would be performed simultaneously in a different section. If conducted in this fashion, with four simultaneous on-site operations, anticipated traffic accessing the site is presented in Table 5.

Once the solar arrays become operational, it is estimated that two full-time personnel would require access to conduct routine operations and maintenance activities. The capacity of the major access routes to the Laboratory, are considered more than adequate to handle the temporary increase in traffic. Once on-site, access to the construction areas are provided through a designated traffic route. The primary site roadways utilized by Laboratory staff and researchers would be avoided.

This route would be the current one used by delivery trucks to the BNL site and is capable of conveying vehicles of this size. Several of the fire-break and access roads located within the project footprint would be eliminated to facilitate and optimize solar array placement. Perimeter fire-break and access roads around the project areas would be maintained to ensure adequate accessibility in the event of a wildland fire or other emergency. Laboratory emergency response personnel would also evaluate all proposed roadway reductions.

Alternative Access Point - Because the proposed action would be located in the south and southeast portion of the Laboratory property, an alternative access point onto the Laboratory site may be considered. The normally locked gate at the southeast corner of the site, near the intersection of the East Firebreak and South Boundary Road, opens onto a local road, North Street. Vehicles using this access point would likely exit the LIE at Exit 69 Wading River Manor Road, proceed north about 1/2 mile, then turn left onto North Street and travel approximately two miles to the southeast access gate. Utilizing this alternative access point would temporarily increase the vehicle traffic on this local road. A security post would be established during times when this gate would be accessed.

Railroad spur - The on-site railroad spur may be available for transporting BP Solar project waste off-site, or for material/equipment delivery. Utilizing the rail line would reduce the number of trucks transporting materials on the local highways. The logistics and practicality of using the railroad would need to be further evaluated by the project staff, and the availability of the spur coordinated with other shipments to or from the BNL site.

Table 5. Anticipated Project Traffic Accessing BNL Site in a 24-Hour Period

A. 50 Passenger Vehicles twice a day
B. Truck Traffic: • Site Clearing: 10 Trucks per day • Structure: 5-10 Trucks per day • Array Module Delivery: 2-3 Trucks per Day
Anticipated traffic is based on 2 delivery trucks with 40 feet (12 meter) trailer beds, and 50 passenger vehicles/ work trucks.

4.7. Cultural Resources

4.7.1. Existing Environment

The *Cultural Resource Management Plan for BNL* (CRMP) [BNL 2005] identifies the Laboratory's historic and cultural resources, and describes the strategies developed to manage them in accordance with applicable laws and regulations. Because most of the proposed project site has seen minimal disturbance since the founding of BNL, there may be a moderate to high potential for the presence of 20th century archeological deposits, including the eras of WWI (1917-1921) and World War II (1940-1946) Camp Upton, and the Civilian Conservation Corps (1934-1939) [Merwin].

The southern area proposed for construction overlays the WWI Remount Facility. This facility was the location where horses and mules were housed prior to deployment of the American Expeditionary Force to Europe and had a number of buildings that housed Remount Facility staff and at least one building that was identified as a YMCA structure. In 2004, the NY State Historic Preservation Officer (SHPO) concurred with the DOE's determination that all WWI period Camp Upton features at BNL, including trenches and foundations, are eligible for listing in the NRHP. In the northern area several out buildings (outhouses) are known to have been present with one still standing in the woods north of the biology field. These features often contain artifacts indicative of the era when they were most used.

4.7.2. Effects of Preferred Alternative on Cultural Resources

BNL performs its cultural resources analyses pursuant to Section 106 of the National Historic Preservation Act. Integrated into the BNL CRMP are recommendations by the Institute for Long Island Archaeology (ILIA) that address the potential for land disturbance/development within the footprint of the former WWI Camp Upton [Bernstein]. Because of the moderate to high potential for the presence of 20th century archeological deposits in the World War I Remount Facility and outhouse area, construction and land preparatory activities in these areas may have some unavoidable adverse effects. Therefore, archeological and data recovery surveys of the WWI Remount Facility and outhouse areas, were performed and documented, as described below. The scope of archaeological work was sufficient to conclude that no further archaeological investigations are recommended.

WWI Remount Facility area: A stage I archeological survey of the proposed project area [approximately 33 acres (13.2 hectares)] was performed by the ILIA in order to assess the overall potential of the area for the presence of cultural resources. This involved both a surface inspection and subsurface testing. The surface inspection entailed a walkover of the entire project area. Subsurface testing involved the excavation of small shovel test pits (STPs) on a closely spaced grid system, in accordance with NY State guidelines.

Privy (outhouse) area: An architectural and archeological data recovery/assessment of the standing Civilian Conservation Corps (CCC) era privy was conducted in accordance with Section 106 of the National Historic Preservation Act. The architectural documentation was designed to meet the standards of the Historic American Buildings Survey (HABS). The archeological survey included excavation of the outhouse pit to document the potential presence of artifacts, and required removal of the privy structure from its rotted sill.

Archaeological Survey Results

WWI Remount Facility - Archival research and archaeological investigation for the location of the proposed BP Solar Array Project indicates that the project area witnessed virtually no discernable use until the early 20th century, when Camp Upton was constructed as a WWI cantonment. Subsurface testing entailed the excavation of shovel test pits in the proposed project area. None of the shovel tests contained an undisturbed soil sequence, reflecting extensive earth-moving activities in the parcel during the 20th century. No prehistoric Native American artifacts or features were encountered. The Stage 1 survey yielded early 20th century Euro-American artifacts (mostly nails, coal, and slag), both probably associated with WWI and/or CCC activities. Two features, WWI Camp Upton concrete building foundations, potentially eligible for listing in the NRHP are located within the proposed project area. The foundations were documented and determined to have no further research potential. While the proposed action would have adverse impact to theses two WWI features, the foundations would be removed, the scope of archaeological work performed was sufficient to conclude that no further archaeological investigations are recommended [Bernstein 2009a].

The extant outhouse structure (privy) was photographed and drawn to scale. There were few artifacts in the privy deposit, and most were architectural debris. Several shovel test pits were made in an effort to identify additional privy deposits that may be present in the area, but none were discovered. Because questions regarding the privy site's age, context, and associations still exist, and because there may be other nearby privy vaults that could yield additional information, cultural resource personnel would monitor the 0.5 acre site if grading/construction work would disturb the soil greater than six inches deep. No further investigations are recommended elsewhere in the parcel [Bernstein 2009b].

Artifacts resulting from the surveys will be appropriately curated and stored as prescribed in the BNL CRMP.

The former BNL Astronomy Club observatory building, constructed in the early to mid-1980s, and the Mixed Waste Handling Building 870, constructed in 1997, are less than 50 years old and have no extraordinary historic significance. Therefore, these structures are not subject to the evaluation requirements delineated in Section 106 of the National Historic Preservation Act.

The current WMF area is located in a highly developed area of the BNL site and has experienced substantial ground-disturbing actions. Archaeological surveys would not be required due to the very low potential for the presence of archeological deposits.

Use of Brownfields

Due to the substantial ground-disturbing actions conducted in the brownfield areas, both during their use as waste disposal sites and during the CERCLA-related remediation, there is a very low potential for the presence of archeological deposits. Therefore, archaeological surveys are not required, and locating PV arrays in these areas would have no adverse effect on any potential cultural resource.

4.8. Air Quality

4.8.1. Existing Environment

The overall regional air quality is affected by a mix of maritime and continental influences. This results in the region, and the BNL site, being very well ventilated by winds from all directions.

The local air quality management in the New Jersey-NY-Connecticut Interstate Air Quality Control Region, which includes Suffolk County and BNL, is in attainment with most National Ambient Air Quality Standards (NAAQS) for criteria pollutants, which include sulfur dioxide, nitrogen oxides, and particulate matter less than 10 microns in diameter (PM_{10}), lead, and carbon monoxide (CO). The region is considered a non-attainment area for ozone. While ozone is a regulated pollutant, it is not emitted directly from sources but is formed by a combination of nitrogen oxides (NO_x) and volatile organic compounds (VOCs) reacting with sunlight in the atmosphere. A NY subset of the region, which includes Kings, Queens, NY, Rockland Westchester, Nassau, and Suffolk counties, is considered a nonattainment area for the 24-hour PM-2.5 (particulate matter less than 2.5 microns in diameter) standard.

4.8.2. Effects of Preferred Alternative on Air Quality

Exhaust from construction, worker, and material delivery vehicles, and other equipment during construction of the proposed site, such as portable electrical generators would result in localized, short-term increases in CO and NOx emissions. Airborne dust ($PM_{2.5}$ and PM_{10} emissions) would also be generated as a result of excavation and vehicle traffic on unpaved surfaces. During construction, fugitive dust generation would be controlled, as needed, by spraying water on soil surfaces and installing stabilized rock construction entrances. Current plans for site grading would maintain the existing grade where possible and leave topsoil in place. Reducing grading actions would minimize particulate emissions. Refer to Section 4.6.2 Transportation Conditions for additional information on the estimated number of vehicles expected. A review of construction operations would be performed to determine if the potential emissions of the project would exceed pollutant thresholds required in 40 CFR Part 93 Subpart B. If a conformity analysis is necessary under the regulation, calculations would be done to determine the potential impact of construction emissions on regional air quality. Any permit (e.g., NYSDEC Air Facility Permit), mitigation or regulatory actions identified as a result of the analysis would be implemented, as necessary.

Once construction is complete, planting native short-growing shade-tolerant grass species for ground cover below the PV arrays would minimize fugitive dust emissions. When operational, the solar PV array facility would not be a source of criteria pollutants. Solar technologies produce near-zero carbon dioxide emissions, which would be a beneficial impact to the regional air quality (Refer to Section 5.9 Climate for additional details).

4.9. Climate

4.9.1. Existing Environment

Climate can influence several environmental parameters including regional and local air quality, storm water drainage, surface waters, and natural hazards.

The climate at the Laboratory can be characterized as breezy and well-ventilated, like most of the eastern seaboard. The Long Island Sound, the Atlantic Ocean, and associated bays influence wind directions and humidity and provide a moderating influence on extreme summer and winter temperatures. The prevailing ground-level winds are from the southwest during the summer, from the northwest during the winter, and about equal from these two directions during the spring and fall [Nagle, 1975; 1978].

BNL has been recording weather data since August 1948 to serve the needs of this DOE site. The average yearly precipitation is 48.75 inches (123.8 centimeters) and the average yearly snowfall is 30.5 inches (77.47 centimeters). The average monthly temperature is 50.2° Fahrenheit (10.1° Celsius). Additional historical meteorological data are available from the BNL Meteorology Services webpage.

Climate Change

In recent years, climate change has evolved into a matter of global concern because it is expected to have widespread, adverse effects on natural resources and systems. A growing body of evidence points to anthropogenic (manmade) sources of greenhouse gases (GHG), such as carbon dioxide (CO_2), as major contributors to climate change. Additional greenhouse gases include methane (CH_4), nitrous oxide (N_2O), halocarbons, and fluorinated compounds. Climate is usually defined as the average weather, over a period ranging from months to many years. Climate change refers to a change in the state of the climate, which is identifiable through changes in the mean and/or the variability of its properties (e.g., temperature or precipitation) over an extended period, typically decades or longer [DOE/EA-1662]. Ongoing climate change research was summarized in reports by the United Nations Intergovernmental Panel on Climate Change (IPCC), *US Climate Change Science Program's Science Synthesis and Assessment Products*, and the *US Global Change Research Program.*

These reports concluded that the climate is already changing; that the change would accelerate; and that man-made GHG emissions, primarily CO_2, are the main source of accelerated climate change [DOE 2009]. Terrestrial carbon sequestration is the process through which CO_2 from the atmosphere is absorbed by trees, plants and crops through photosynthesis, and stored as carbon in biomass (tree trunks, branches, foliage and roots) and soils. Forests and soils have a large influence on atmospheric levels of CO_2, essentially helping to mitigate man-made CO_2 emissions [EPA 2006].

Various GHGs differ in their potential contribution to global warming. The global warming potential (GWP) compares the relative ability of each GHG to trap heat in the atmosphere over a certain period. According to guidelines, CO_2 is the reference gas with a GWP of 1. Based on a period of 100-years, the GWP of methane is 21, implying that a ton of methane is 21 times more effective in trapping heat than a ton of CO_2. The GWP for N_2O is 310. Carbon dioxide equivalent is a measure that expresses, for a given mixture and amount of greenhouse gas, the amount of CO_2 that would have the same GWP [Haile].

4.9.2. Effects of Preferred Alternative on Climate

Current methodology is not able to directly correlate GHG emissions from discrete projects to any specific impact on climate change. However, constructing and operating a large-scale PV solar array on the BNL site may affect GHG emissions in multiple ways. The temporary increase in vehicle exhaust emissions during the project construction phase, described in Section 5.8 Air Quality, may provide minimal contribution to increased GHG emissions.

Removing approximately 153 acres (52 hectares) of trees, out of some 3,607 acres (1,460 hectares) of on-site forested area, as part of the proposed action would eliminate a carbon absorption/storage source that is estimated to sequestrate an average of about 5.5 metric tons CO_2 equivalent per acre each year (13.6 metric tons CO_2 equivalent/hectare/year) or 842 metric tons of CO_2 equivalent each year for the entire stand of trees. Over 40-years, the estimated life of the solar arrays, the ability to sequestrate about 33,680 metric tons of CO_2 equivalent would not be realized. In comparison, the net sequestration in all US forests, urban trees and agriculture in 2001 totaled an estimated 840 million metric tons of CO_2 equivalent [EPA 2006]. Removing a carbon storage source of this size would not be expected to have a direct impact on climate change.

In comparison, the proposed 37 MW solar arrays would have a beneficial savings effect on GHG emissions by producing electricity with near-zero carbon dioxide emissions. If conventional means such as burning fossil fuels were used to generate an equivalent amount of electricity, about 30,950 metric tons of CO_2 per year would be emitted into the atmosphere.

Additional GHG and criteria pollutants discharged by conventional generating methods are presented in the table below:

Table 6. EPA eGrid Output Emissions Rates for Long Island Region [eGrid]

	Emissions Avoided for a 37 MW Project
Greenhouse Gases	**(CO_2 equivalent metric tons /yr)**
Carbon Dioxide (CO_2)	30,950
Methane (CH_4)	2
Nitrous Oxide (N_2O)	0
Criteria Pollutants	**(metric tons/yr)**
Nitrogen Oxides (NOx)	33
Ozone Season NOx (May-Sept)	30
Sulfur Dioxide (SO2)	76

Over an estimated 40-year life of the project, the discharge of about 1,238,015 metric tons of CO_2 into the atmosphere from conventional means would be avoided. Considering the positive impact of the project lifetime CO_2 emissions avoidance (1,238,015 metric tons of CO_2 equivalent), along with the adverse loss of sequestration volume (33,680 metric tons of CO_2 equivalent), indicates that the proposed action would have an overall beneficial effect on GHG emissions and climate.

4.10. Visual Quality

4.10.1. Existing Environment

Large scientific facilities and structures have been constructed and operated at BNL since the late 1940s. Such structures have included research reactors with a 310-foot (94.5 meter) exhaust stack located on the highest point of the BNL site and a 100- foot (30.5 meter) tall meteorological tower. Current visual features of the proposed project area consist primarily of a wooded pine and oak landscape, with a mix of open fields located in the north array section.

A small cluster of four or five residential homes are situated immediately south and within 200-500 feet (61-152 meters) of the LIRR track that borders the Laboratory's southern boundary.

From these properties, the current view to the northwest includes narrow wooded buffers, less than 10 feet (about 3 meters) wide, on each side of the tracks. Seasonal views, from November to April when the trees are bare, include the southeastern edge of wooded area proposed for the south array.

Views From LIRR Looking North.

Views From North Street Looking North.

Use of Brownfields

Current visual features associated with the brownfield areas encompass an open landscape surrounded by woods. The open areas present views of grasses, one to three feet (0.3 -1 meter) high, covering the Former Landfill and Interim Landfill mounds, with the remaining areas essentially barren or with minimal vegetation. An exception to this view is the NSLS-II construction-site that is visible to the northwest.

4.10.2. Effects of Preferred Alternative on Visual Quality

The surface elevation of proposed locations for the BP solar arrays would be equal to or slightly less than that of BNL's core developed area. The solar arrays would be constructed to a maximum height of about 11 feet (3.3 meters). Because the proposed location is in the central south and southeast portion of the Laboratory property, and approximately three-quarters to one mile (1,207 to 1,609 meters) away, the majority of arrays would not be visible from the Laboratory's central developed area. If solar arrays were situated on the former landfill area, they may be visible from the future NSLS-II facility and associated grounds.

NY Department of Environmental Conservation (DEC) Guidance "Assessing and Mitigating Visual Impacts" (DEP-00-2, Issuance Date: July 31, 2000) states that:

> "Aesthetic impact occurs when there is a detrimental effect on the perceived beauty of a place or structure. Significant aesthetic impacts are those that may cause a diminishment of the public enjoyment and appreciation of an inventoried resource, or one that impairs the character or quality of such a place."

A visual analysis was conducted in a manner consistent with the referenced DEC Guidance. Out of the 15 aesthetic resource categories listed in the DEC guidance, two inventoried resources, located on the Laboratory property, were evaluated for potential visual impacts from the proposed action. – (Associated DEC aesthetic resource category is identified in parentheses): The WW I Camp Upton Training Trenches (A property on or eligible for inclusion in the National or State Register of Historic Places); and the Peconic River (Rivers designated as National or State Wild, Scenic or Recreational).

At its closest point, the proposed solar arrays would be approximately 2,200 feet (671 meters) from the nearest Camp Upton training trench, and 2,000 feet (610 meters) from the Peconic River. However, due to the density of trees separating the trenches and river from the project area, the panels would not be seen from either aesthetic resource.

Off-Site

The majority of the proposed solar arrays would not be seen from off-site due to their low vertical profile and the extensive wooded buffer surrounding the Laboratory property. However, the southern edge of the south array would be located adjacent to the Laboratory's southern boundary and immediately north of the LIRR Ronkonkoma to Greenport line tracks. This group of arrays would be visible from passing trains for approximately 1,100 feet (335 meters). Ridership on this line is relatively low. Based on the current LIRR schedules, no more than eight trains travel by the site weekdays, and no more than four trains on weekends [LIRR]. Each train usually has two to four train cars per trip.

The same group of arrays would also be visible from a few of the residential properties situated immediately south of the LIRR tracks, though primarily from November to April when the trees are bare. The nearest public road to the proposed project area is North Street, which runs southwest to northeast and at its closest point would be approximately 600 feet (183 meters) from the southeast corner of the south array. The combination of roadway elevation and existing trees within the residential properties and LIRR buffers would make it difficult for the PV arrays to be seen from a vehicle traveling along North Street. Without leaves on the trees it may be possible to see the southeast corner of arrays for an estimated distance along North Street of about 200-300 feet (61-91 meters). No known inventoried aesthetic resources are located off-site within the potential visual field of the proposed solar arrays.

The proposed action would be visible from aircraft flying over or near the Laboratory and via satellite. The solar array design would include an anti-reflective coating that increases the light transmittance by reducing reflectance at the glass surface. This coating may also reduce any potential reflective glare visible from the arrays. Based on the increasing use of PV arrays at commercial and military airports around the world, reflective glare would not be expected to have an adverse impact on aircraft.

Overall the Proposed Action would not be expected to have an adverse visual impact either on or off the Laboratory property.

4.11. Noise

Noise is defined as unwanted sound that interferes with normal activities or in some way reduces the quality of the environment. Response to noise varies according to its type, perceived importance, appropriateness in the setting and time of day, and the sensitivity of the individual receptor. The EPA developed an index (threshold) to assess noise impacts from a variety of sources using residential receptors. If daytime noise values exceed 65 decibels (dBA), residential development is not recommended [EPA 1974]. Noise sensitive receptors are defined as the occupants of a facility or a location where a state of quietness is a basis for use or where excessive noise interferes with the normal use of the facility or location. Typical noise sensitive receptors include schools, hospitals, churches, libraries, homes, parks, and wilderness areas.

4.11.1. Existing Environment

The majority of the proposed project area, located more than one mile (1.6 km) from the core developed sector of the Laboratory, is surrounded primarily by woodlands. There is a wide range of existing noise sources present in the area of the proposed solar array. The types of sources that contribute to ambient noise levels may include: street traffic such as cars, trucks, buses, rail traffic, aircraft over-flights, commercial and industrial noise sources, noise from existing power lines, residential noise sources, and rural environment sources (wildlife, etc.).

On-site noise sensitive receptors may include the occasional BNL employee jogging, biking, walking or performing work, such as well-drilling or environmental sampling. Two potentially sensitive receptors were identified with respect to noise impacting a horse farm just south of the site, across the LIRR track and adjacent to the LIPA substation, and a residence along North Street about 3,000 feet (914.4 meters) east of the northeastern corner of the south array [ERM].

The closest home is roughly 250 feet (76.2 meters) from the southern edge of the project site, with other houses located about 800 feet (244 meters) way. At its closest point, the project would be about 2,000 feet (610 meters) from the Peconic River. Public access is not permitted on this section of the river located on DOE property. Private homes and public lands in the vicinity, located approximately 1.5 miles (2.4 km) from the project area, may also be considered noise sensitive receptors.

For a more rigorous assessment of the impacts of site-generated noise on local residents, a noise monitoring survey was conducted for two days in July 2009 at the horse farm located adjacent to the BNL site. A sound level meter was placed on a tripod near the northern fence line of the horse farm, at its closest point to the proposed construction activities. During daytime periods (7 am to 7 pm), the median sound reading at the site was 51.6 dBA. The highest noise readings with no time constant applied were as high as 113 dBA, and the maximum sound level during a time-weighted period was 87.5 dBA – both were measured when the LIRR trains pass by.

4.11.2. Effects of Preferred Alternative on Noise

Construction activities associated with the proposed solar arrays would result in temporary increases in ambient noise levels for approximately 16 months. A variety of construction equipment such as tree harvesters, graders, dozers, trenchers, pile drivers, dump trucks, and delivery trucks would generate noise intermittently during daylight hours. Noise levels from construction-sites measured approximately 90 dBA at a distance of 50 feet (15.24 meters) from the center of the site [CERL]. Sites in flat-lying areas with minimal vegetation experience noise attenuation at a rate of 6 dBA for each doubling of distance between the source and the receptor [CERL]. A receptor located one mile (1.6 km) away from the proposed site would hear noise levels at approximately 45 dBA and therefore would not be negatively impacted by construction activities. A receptor located 200-800 feet (61 - 244 meters) from the construction-site would experience noise levels at approximately 78-66 dBA, respectively. Noise levels at the Peconic River's closest point would be about 60 dBA. These estimated noise levels are considered conservative values because the proposed project site is generally surrounded by woodlands, thereby attenuating the levels to a greater degree, resulting in lower noise levels. The loudest noise levels anticipated for the 50-60 percussive pile drivings per day would be around about 101 dBA, 50 feet (15 meters) away from the source. The levels would attenuate to about 89 dBA at the property line of the horse farm, and less than 67 dBA at the closest North Street residence.

For comparison noise levels are commonly compared to typical noise sources encountered in public are shown in Table 7.

Table 7. Common Noise Exposures

Sound Source	Pressure Decibels dBA
Large rocket engine (nearby)	180
Jet takeoff (nearby)	150
Pneumatic riveter	130
Jet takeoff (200 feet)	120
Construction noise (10 feet)	110
Subway train (100 feet)	100
Heavy truck (50 feet)	90
Average factory	80
Normal conversation (3 feet)	60
Quiet office	50
Library	40
Soft whisper (16 feet)	30
Rustling leaves	20
Normal breathing	10
Hearing threshold	0

NY Department of Environmental Conservation (DEC) Guidance "Assessing and Mitigating Noise Impacts" (DEP-00-1, Issuance Date: October 6, 2000 Revised: February 2, 2001) states that:

> "Increases ranging from 0-3 dBA should have no appreciable effect on receptors. Increases from 3-6 dBA may have potential for adverse noise impact only in cases where the most sensitive of receptors are present. Sound pressure increases of more than 6 dBA may require a closer analysis of impact potential depending on existing Sound Pressure Level (SPLs) and the character of surrounding land use and receptors."

As can be seen from a preliminary assessment of the data, peak sound levels at the fence line of the horse farm predicted even during pile driving activities (89 dBA) would be substantially less than peak sound levels already experienced at the site, and would represent less than a 6 dBA increase over background levels (87.5 dBA) experienced when the LIRR trains pass by. As construction moves to other portions of the site, the noise generated during pile driving would quickly fall to less than peak background noise – at 500 ft. (152 meters) the noise resulting from pile driving would be 81 dBA.

Common noise sources during construction would include grading, bulldozing, and truck loading and unloading. At 200 feet (61 meters), these would generate noise impacts of approximately 70 dBA, which would fall to 65 dBA at 500 feet (152 meters) and 50 dBA at 2,500 feet (762 meters). The average background at the horse fence line was approximately 51.5 dBA. Thus, when construction work is closer than approximately 2,000 feet (610 meters), it would result in a greater than 6 dBA increase in noise over background.

During worst-case times (highest noise activities, activities taking place near to the southern or eastern boundaries of the project site), there would be occasions when the NYSDEC guidance of 6 dBA is exceeded. However, these situations would not have a major adverse effect because the construction phase would be limited to 16 months, and only during a small fraction of those 16 months would the highest noise generating activities be taking place at the southern or eastern-most portions of the site.

Heavy equipment would generate noise that could affect the project-site workers during construction. Construction equipment typically emits noise in the 85 dBA to 135 dBA range. Laboratory safety programs and the construction contractor would require workers to wear hearing protection in accordance with OSHA regulations. Operational noise from the solar panel arrays and associated maintenance activities would be negligible and would likely be inaudible against ambient levels.

4.12. Industrial Safety and Occupational Health

4.12.1. Existing Environment

The area proposed for the PV arrays is currently undeveloped, consisting of woods, open fields and unpaved roads. As a result, the predominant industrial safety and occupational health (IS&H) concern is motor vehicle accidents during inclement weather (ice or heavy rains), sprains, strains, and falls sustained by employees while exercising, and insect bites during environmental field operations such as sampling and well drilling. The Laboratory maintains an Occupational Medical Clinic staffed with doctors and nurses to evaluate and treat non-emergency injuries, as well as an extensive emergency management program that encompasses planning for and response to accident events. The on-site Fire-Rescue Group includes trained Emergency Medical Technicians (EMTs) and response vehicles.

4.12.2. Effects of Preferred Alternative on Industrial Safety and Occupational Health

Construction and trade workers would be exposed to safety and health hazards faced at similar construction sites. Potential impacts to the health and safety of the workers would be minimized by adherence to federal, state, and local regulations, OSHA regulations, and general contractor safety plans. Electrical work would conform to applicable electrical and fire code requirements. No unusual construction site considerations are expected during the installation and maintenance of the proposed solar power arrays and associated equipment.

BNL employees and the general public would not be adversely impacted by the construction and operation of the proposed solar array project. The remote location of the site and construction of a fenced enclosure would minimize exposure of the Laboratory staff, visitors, and public to potential safety hazards at the site.

Use of Brownfields

The primary residual chemical surface contaminant of concern in the Chemical/Animal Pits and Glass Holes Area is mercury. Residual mercury levels in the soils were remediated to meet industrial standards. Locating solar arrays in this area would be considered within the industrial uses identified in the BNL Land Use and Institutional Control Program. The Glass Holes Area is currently being used as a site composting operation. Reference Section 5.13.1 regarding the potential use of capped landfill areas.

4.13. Radiological Characteristics

4.13.1. Existing Environment

The radiological characteristics of the BNL site are presented below as an overview of Laboratory's monitoring efforts in order to provide a baseline for potential environmental effects.

The radiological characteristics of Laboratory operations are determined through routine and permit-based monitoring efforts. Water discharged from the STP is routinely monitored at the plant's Peconic River Outfall. In 2007, all effluents were found to be less than the Safe Drinking Water Act limits of 4 millirem annual dose limit for gross beta, 15 picocuries per liter (pCi/L) for average gross alpha activity, and 20,000 pCi/L average tritium concentration.

BNL uses 10 recharge basins permitted under SPDES to discharge once-through cooling water, cooling tower blow-down, and storm water runoff. Routine monitoring of these basins indicated that the average concentrations of gross alpha and beta activity were within typical ranges and that there were no gamma-emitting radionuclides detected. In 2007, there was a single, low detection of tritium in the discharge to Recharge Basin HT-W, which receives once-through cooling water and cooling tower blow down. The maximum concentration detected was 430 pCi/L, which is approximately 2 percent of the drinking water standard. [SER]. There were no radionuclides detected in 2008.

BNL is subject to the requirements of 40 CFR Part 61, Subpart H National Emission Standards for Hazardous Air Pollutants (NESHAP). The U.S. EPA established a national policy on the airborne emission of radionuclides, and a dose limit to the public of 10millirem/yr for the airborne pathway. The effective dose equivalent from all air emission sources at BNL for 2007 was calculated to be 0.06millirem, far below the allowable limit [SER].

The Former Hazardous Waste Management Facility (FHWMF) was cleaned up in 2005 to industrial standards. Areas to the northeast and southeast of this area are proposed for development. Recent monitoring has resulted in additional clean up of small areas outside of the FHWMF in 2009, with potential additional clean-up being necessary.

The radiological characteristics of groundwater contamination plumes in the vicinity of the proposed action are discussed in Section 4.3 Water.

Use of Brownfields

The Chemical/Animal Pits and Glass Holes Area have been remediated to meet industrial cleanup goals. While remaining radioactivity in the soil includes a mixture of Cesium-137 (Cs-137) and Strontium-90 (Sr-90), current soil conditions meet worker exposure guidelines for radioactivity. The glass holes area is currently being used as a site composting operation.

The landfill areas have been capped to prevent site workers, employees, visitors, and wildlife from being exposed to the subsurface contamination. Contaminated soils and materials are present below these capped areas. These landfills are currently monitored and maintained by mowing the grass, keeping the cap intact and monitoring landfill gas and groundwater.

4.13.2. Effects of Preferred Alternative on Radiological Characteristics

As presented in Section 4.3 Water, constructing solar arrays in the proposed locations would not affect the hydrologic or radiological characteristics of the groundwater plumes. Due to the plume depths and proposed construction methods, there would be no potential for worker exposure during construction, operational or decommissioning activities.

Use of Brownfields

Solar arrays located in capped Land Fill Areas would be anchored in place using concrete ballasts to prevent damage or penetration of the protective cap and potential exposure to radiological and chemical contaminants. Based on the Chemical/Animal Pits and Glass Holes Area cleanup level to industrial standards, there would be minimal risk of worker exposure to contaminants during construction, operational or decommissioning activities. In addition, radiological and work control measures would be utilized to further reduce any potential risk. Examples of control measures include, but are not limited to, adding additional soil layers, administrative controls on soil disturbance at specific depths, restrictions on in-ground pile removal, and radiological monitoring.

4.14. Natural Hazards

4.14.1. Existing Environment

Natural phenomena, which could lead to operational emergencies at BNL, include hurricanes, tornadoes, wildfires, thunderstorms, snowstorms, and ice storms. Hurricanes occasionally hit Long Island and the high wind speeds associated with them may potentially damage structures. Record high winds for BNL were recorded during Hurricane Carol in September 1954 [Hoey]. Tornadoes and hailstorms are extremely rare on Long Island. Thunderstorms, snowstorms, and ice storms do occasionally occur and have the potential to cause damage to facilities.

The banks of the Peconic River, which traverse portions of the eastern side of the BNL site, are within the Federal Emergency Management Agency (FEMA) designated 100-year floodplain [FEMA].

Earthquakes on Long Island are extremely rare, and no active earthquake-producing faults are known in the Long Island area [Hoey]. Long Island lies in a zone 2 or moderate damage seismic probability area and it is assumed that an earthquake of Modified Mercalli VII could occur [HFBR EIS]. A recent history of earthquakes in the central Long Island area is presented below [USGS]:

Table 8. Recent History of Earthquakes in the Central Long Island Area

Year	Date	Intensity - Modified Mercalli
1925	Feb 25	I-III
1929	Nov 18	I-III
1935	Nov 1	I-III
1937	Jul 18	I-III
1944	Sep 5	I-III
1950	Mar 29	I-III
1951	Jan 25	I-III
1985	Oct 19	IV-V (4-5 on Richter scale)

The likelihood of a serious earthquake in the BNL area is slight and seismologists expect no significant earthquakes in the foreseeable future [Hoey].

The Central Pine Barrens and community types within BNL are fire dependent systems that experience periodic wildfire events. Wildfires, direct flame and smoke could affect BNL operations. The BNL Wildland Fire Management Plan (WFMP) includes recommendations for periodic mechanical tree thinning and prescribed fire (controlled burns) to reduce potential fuel loading and the effects of unanticipated wildfire ignitions [BNL 2003b]. Prescribed burns, totaling about 16 acres (6.5 hectares), have been performed in two out of the last five years. The WFMP also recommends that a cleared area of at least 30 feet (9 meters) be maintained between buildings and the nearest treed area. The BNL on-site fire department is manned 24-hours a day to respond to all fire emergencies, and maintains mutual aid agreements with local fire departments.

4.14.2. Effects of Natural Hazards on Preferred Alternative

At the closest point, the proposed solar arrays would be approximately 2,000 feet (610 meters) from the Peconic River and 10 feet (3 meters) higher in elevation. The potential may exist for some seasonal flooding in the north array area due to its proximity (approximately 100 feet or 30.5 meters) to wetlands. However, damage to the arrays would not be expected because the arrays would be supported at a minimum of 2 feet (0.6 meters) off the ground.

The solar arrays and associated electrical infrastructure would be designed and built to applicable codes and standards, including the ability to withstand

120 mph (193 kph) hurricane winds. Structure failure of an array due to natural phenomenon would result in physical damage to the array. The solar arrays do not contain any hazardous materials, so there would be no fluids to cause insult to personnel or the environment. Because the arrays would be located away from densely populated areas of the Laboratory site and the surrounding community, human health and safety issues (i.e., injuries) would likely have a low probability of occurring as a result of structural failure.

The potential for wildfires to affect and damage solar panels and associated electrical equipment would be similar to other BNL structures. Clearance between the arrays and the nearest forested/vegetation area would reduce the potential for fire spread. Existing BNL wildfire control and response measures identified above would also apply to the solar area project areas.

4.15. Intentional Destructive Acts

Construction and operation of the proposed large-scale solar photovoltaic array project on the BNL site would not involve the transportation, storage, or use of radioactive, explosive, or toxic materials. Consequently, it is highly unlikely that saboteurs or terrorists would view construction or operation of the arrays as a potential target. The project location is not near any national defense infrastructure or in the immediate vicinity of a major inland port, container terminal, freight trains, or nuclear power plants. In addition, fencing and random patrols by the Laboratory Police and inspections by vendor personnel mitigate the potential for destructive acts. The Proposed Action would not offer any targets of opportunity for terrorists or saboteurs to inflict adverse impacts to human life, heath, or safety.

4.16. Utilities

4.16.1. Existing Environment

Few utilities are present within the footprint of the proposed project site because these areas have remained largely undeveloped since the inception of BNL in 1947, with the exception of some experimental agricultural programs. A natural gas line enters the site above ground at the southeast gate off North Street, proceeds underground along the south side of the South Boundary Road utility easement, then runs northwest along Powerline Road to the steam plant.

As part of an existing utility easement, overhead 13.8 kV high voltage power lines run along the following firebreak roads bordering the proposed solar array sites: Fifth Avenue, East Firebreak, South Boundary Road, and Powerline Road. Other electric lines near the project areas include:

- 2400 Volt line running underground along the firebreak roads north and west sides of the South Array.
- Abandoned underground 480 Volt line along the firebreak road north of the South Array.
- Control wiring about 2.5 feet (0.76 meters) below grade along Powerline Road

Abandoned terra cotta sewer lines and circular brick manholes, dating from WWI Camp Upton are located in the proposed project area that overlay the WWI Remount Facility.

An existing electrical substation (8ER), operated by LIPA, is located south of and adjacent to the LIRR tracks located along the Laboratory's south border.

4.16.2. Effects of Preferred Alternative on Utilities

As presented in Section 4.1, the intent of the proposed BP solar array project is to provide 37 MW of power to the LIPA electric grid system. During periods of peak electrical demand, typically middle of day, hottest period, operation of the proposed BP solar arrays would result in a reduced need for extra energy generated through conventional means. The electrical configuration is detailed in Section 4.1. Power from the 37 MW BP solar arrays would be connected to the grid via the LIPA substation located adjacent to the Laboratory's south border. Three conduit lines, one for fiber optics and two for power distribution (one active and one spare), would be run under the LIRR tracks to connect the transmission line from the BP solar array 13.8/69kV transformer to the substation. Refer to Section 5.19.1 for further details.

The proposed Laboratory dedicated array may provide 1 to 2 MW of solar-generated power to the Laboratory's internal electrical grid. Depending on its final location, either the Chemical/Animal Pits and Glass Holes Area or current WMF yard, a transmission line would be run from a transformer/inverter pad to a BNL substation located on Cornell Avenue, or near East Fifth Avenue. Water, sanitary, and electric utilities would be run from existing lines to a new dedicated array support building. Renovating

Building 870 to serve as the support building would utilize existing utilities. Newly routed lines would be located in previously disturbed areas.

Temporary trenching for utility line installation would be returned to grade following installation.

All existing modern underground utility lines would be located and marked prior to initiating construction actions. If any abandoned WWI sewer lines and manholes are encountered during construction, they would be removed.

4.17. Electric and Magnetic Fields (EMF)

There are no Federal standards limiting residential or occupational exposure to the common utility magnetic or electric fields found in the United States. The applicable electric field strength standards established by the New York State Public Service Commission (PSC) are set forth in the Opinion No. 78-13, issued June 19, 1978. The magnetic field standards are set forth in the PSC's Interim Policy Statement on Magnetic Fields, issued September 11, 1990.

Opinion 78-13 established an electric field strength interim standard of 0.5 kilovolts per foot (1.6 kilovolts per meter (kV/m)) for electric transmission lines, at the edge of the right-of- way, 3.3 feet (1 meter) above ground level, with the line at the rated voltage. The Interim Policy established a magnetic field strength interim standard of 200 milligauss (mG), measured at 3.3 feet (1 meter) above ground grade, at the edge of the right-of-way, at the point of lowest conductor sag [Caithness].

4.17.1. Existing Environment

Overhead high-voltage electric power lines are currently present in the vicinity of the proposed action, including a 13.8 kV LIPA line running along an existing utility easement to LIPA substation 8ER, and a 69kV line from the substation 8ER to the BNL onsite substation. The power lines were constructed and are maintained according to applicable requirements. There has been no indication to date of any environmental effects from EMF associated with these lines.

4.17.2. Effects of Preferred Alternative on EMF

Engineering design of the proposed project thus far has not been completed in sufficient detail to provide specific EMF levels. However, electric and magnetic fields associated with the proposed action would be significantly below the 200 mG and 0.5 kilovolts per foot (1.6 kV/m) guidelines, and would not be expected to have any adverse health effects.

4.18. Waste Management and Pollution Prevention

4.18.1. Existing Environment

The Laboratory has implemented extensive and active pollution prevention (P2) and recycling programs that reflect the national and DOE P2 goals and policies. The Laboratory's EPD is staffed with subject matter experts responsible for evaluating and implementing regulatory requirements and P2 programs. The EPD operates the Waste Management Facility (Buildings 855 and 860) where waste generated at BNL is processed and prepared for off-site shipment and disposal. Additional details of the P2 and recycling programs are described in Chapter 2 of the Site Environmental Report [SER].

4.18.2. Effects of Preferred Alternative on Waste Management and Pollution Prevention

Waste products resulting from construction activities would include cleared trees and brush, concrete and steel debris from obsolete structural features.

Trees suitable for timber would be identified and options evaluated to recycle or gain maximum benefit from the trees. Options may include, but are not limited to, shipping the wood to an energy recovery facility, and working with companies that produce wood products directly from supplied trees. Trees unsuitable for timber harvest and remaining tree material, such as limbs, tree tops, etc., would be mulched on-site or composted for topsoil.

If the construction contractor would need to maintain a temporary fuel storage tank on-site for refueling construction vehicles, the facility is required to conform to the requirements of Suffolk County Department of Health – Article 12, Toxic and Hazardous Materials Storage and Handling Controls [Article 12].

The principal goal of the proposed action is to provide clean, renewable electricity produced by solar PV arrays to LIPA. Providing solar generated electricity to the grid lessens the demand for electricity produced though conventional means, such as fossil fuel burning plants, thereby lowering airborne emissions. Section 5.8 Air Quality presents additional detail on the effects of the proposed action on air emissions.

4.19. Commitment of Resources

Construction

Construction activities would result in a temporary increase in fuel use to power the construction vehicles and an estimated 2,000 gallons of water per day for dust control.

Operations

Operation of the proposed BP solar arrays would advance the DOE's vital mission of meeting the nation's energy goals from clean, renewable energy sources and result in a reduced need on the LIPA grid for electricity generated through conventional means. Overall, there would be reductions in the demand for fossil fuels and water resources used to support these conventional power generating methods. It is estimated that as much as 500,000 gallons of water per year, representing a 0.1 percent increase in current BNL water usage, would be required for cleaning PV panels.

4.20. Connected Actions

4.20.1. LIPA Interconnection at Substation 8ER

Because below ground horizontal directional drilling would be used to construct the transmission line from the BNL properly line to LIPA Substation 8ER, the removal of any vegetation would not be required and there would be no impacts to surface conditions. The short route also would not cross any wetlands or any special habitat areas for threatened or endangered species. Additionally, water for horizontal directional drilling mud would be obtained from existing BNL water supplies or trucked-in sources.

The modifications to the substation and the work to construct the transmission line would involve approximately 20 vehicle trips per day (or less) over a one-month period. LIPA would complete an Environmental Assessment Form (EAF) under the NY State Environmental Quality Review Act (SEQRA). Accordingly, the construction of the transmission cable between the BNL property line to the LIPA Substation 8ER and the minor modifications to the LIPA substation would not result in any environmental impacts other than those that have already been disclosed in the EA and EAF, and, overall, would not result in any significant adverse impacts to the environment.

4.20.2. Work for Others

As consideration in the easement agreement between DOE and BP Solar for allowing BP access and use of the federal BNL property, BP Solar may provide funding to the Laboratory to support a Work For Others (WFO) program in research and development of solar technology. The WFO program would allow BSA to perform work for BP Solar, a non-DOE entity, on a fully reimbursable basis. WFO projects utilizing the Laboratory dedicated array are included within the scope of this EA.

4.20.3. Tiger Salamander Breeding Pond Enhancement

Proposed improvements to the area surrounding the subject Tiger Salamander breeding pond would enhance the immediate habitat by enabling persistent water retention. Larval salamanders in the pond would subsequently be able to develop to maturity.

Enhancement actions would be scheduled so that the work would not occur during salamander breeding/foraging periods. In addition, the BNL digging permit process would alert construction personnel to tiger salamander awareness and reporting protocols. A wetlands permit would be obtained from the NYSDEC for the enhancement actions.

4.21. Future Upgrades

While no specific upgrades to the proposed solar arrays or project footprint are planned at this time, it is reasonable to expect that technological advances may warrant changing some or all PV panels or associated equipment in order to achieve improved efficiency. Replacing or upgrading panels on the same or similar support structures, located within the current proposed footprint, would not be expected to have any adverse environmental impacts. Future upgrades that would expand the arrays beyond the current proposed footprint would be reviewed through established NEPA procedure.

4.22. Decommissioning and Restoration

Decommissioning activities would be performed as part of a planned process. It is reasonable to assume that most of the steel, fencing, and hardware would be recycled. After restoration to its pre-project condition, minus expected wear and tear, the land would then be available for future BNL operations or remain open for restoration and natural re-growth. Restoration actions, including planting native pine barrens species, would be developed and implemented in accordance with the BNL Natural Resource Management Plan [BNL 2003a].

4.23. Cumulative Impacts of the Proposed Action

Beyond a temporary increase in vehicle fuel usage during the construction phase, the proposed action would require either none or minimal fuel or water resources. When considered along with previously planned and evaluated actions at BNL, the cumulative impacts would have a negligible effect on the environment.

Reasonably foreseeable projects planned for the Laboratory site are estimated to require the removal of about 206 acres (83 hectares) of trees as shown in Table 8 below.

The total amount of tree removal required for the anticipated BNL projects identified above would amount to approximately 3.7% of the BNL property. This would increase the cleared area of the BNL site from about 26.8% to 30.5%, which is well within the Central Pine Barrens Land Use Plan guidelines.

Table 9. Estimated Tree Removal for Future BNL Projects

Project	Estimated Tree Removal Acres (hectares)
National Synchrotron Light Source-II	8 acres (3.2 hectares)
Interdisciplinary Science Buildings I and II	None
Relativistic Heavy Ion Collider II (RHIC-II)	<5 acres (2 hectares)
e-RHIC	25-35 acres (10-14 hectares)
Revised Main Gate Entrance Road	2-5 acres (1-2 hectares)
BP Solar Array	153 acres (62 hectares)
TOTAL	206 acres (83 hectares)

Overall the cumulative impact of these environmental aspects would have negligible effects on the environment.

4.24. Environmental Consequences of the No Action Alternative

Under the No Action Alternative, DOE would not provide an easement for construction and operation of a proposed 37 MW solar array to produce energy.

Land use and demography on-site and off-site would remain in its current configuration under the No Action Alternative.

The No Action Alternative would have minimal impact on the ecology at the BNL site. These minimal impacts would result from the continued encroachment of invasive species that would likely occur.

By not performing the proposed action, the Laboratory, and local/regional economies would not realize the benefits gained from direct power purchasing, an increased demand for labor, or related secondary benefits. Similarly, the LIPA electric grid would not receive 37 MW of 'clean' renewable electricity produced by solar PV arrays. The 37 MW demand would then be satisfied with electricity generated through conventional means such as fossil fuel plants, along with the associated airborne emissions, and demand for fuels, water, and other resources.

5.0. Acronyms, Initials, and Abbreviations

ALARA	As Low As Reasonably Achievable
AMSL	Above Mean Sea Level
BER	Brookhaven Executive Roundtable
BHSO	Brookhaven Site Office (DOE)
BNL	Brookhaven National Laboratory
BSA	Brookhaven Science Associates, LLC.
CAC	Community Advisory Council
CAP	Clean Air Assessment Package
CEQ	Council on Environmental Quality
CFR	Code of Federal Regulations
CRMP	Cultural Resource Management Plan
dBA	Decibel
DEC	Department of Environmental Conservation

DOE	Department of Energy
EA	Environmental Assessment
EIS	Environmental Impact Statement
EMF	Electric and Magnetic Fields
EMS	Environmental Management System
EMT	Emergency Medical Technician
EPA	Environmental Protection Agency
EPD	Environmental Protection Division
ESH	Environment, Safety and Health
FEMA	Federal Emergency Management Agency
FONSI	Finding of No Significant Impact
FY	Fiscal Year
GeV	Giga-[Billion] Electron Volt
gpd	Gallons Per Day
ILIA	Institute of Long Island Archaeology
IR	Infrared
ISM	Integrated Safety Management
ISO	International Organization for Standardization
kV	Kilovolt
kVA	Kilo volt-amp
kW	Kilowatt
lbs	Pounds
LIPA	Long Island Power Authority
LIRR	Long Island Railroad
MEI	Maximally Exposed Individual
meV	Milli-[thousandth] electron Volt
mG	milligauss
mgd	Million gallons per day
mrem	Milliroentgen equivalent man (see below for "rem")
MW	Megawatt
MWH	Megawatt Hours
NAAQS	National Ambient Air Quality Standards
NEPA	National Environmental Policy Act
NESHAP	National Emission Standards for Hazardous Air Pollutants
NFPA	National Fire Protection Act
NRHP	National Register of Historic Places
NRMP	National Resource Management Plan
NSLS-II	National Synchrotron Light Source-II

NY	New York
NYS	New York State
NYSDEC	New York State Department of Environmental Conservation
NYSPDES	New York State Pollutant Discharge Elimination System
OHSAS	Occupational Health and Safety Assessment Series
OSHA	Occupational Safety and Health Administration
P2	Pollution Prevention
pH	Activity of Hydrogen Atoms in Solution
pCi/l	Pico-[trillionths] Curies per liter [Curie = basic unit used to describe the intensity of radioactivity in a sample of material]
rem	Roentgen equivalent man [standard unit that measures the effects of ionizing radiation on humans]
SBMS	Standards Based Management System
SC	Suffolk County
SER	Site Environmental Report
SEQRA	New York State Environmental Quality Review Act
SPDES	State Pollutant Discharge Elimination System
SPL	Sound Pressure Level
STP	Shovel Test Pit
TS	Tiger Salamander
U.S.	United States
USC	United States Code
USGS	Untied States Geological Survey
VOC	Volatile Organic Compound
WFMP	Wildland Fire Management Plan
WWI	World War I
WSRRA	Wild, Scenic, and Recreational Rivers Act
yr	Year

6.0. LIST OF PREPARERS

Irene Atney, Counsel, DOE/SC-BHSO
Mark. Davis, NEPA Coordinator, EPD, BSA
George Goode, Division Manager, Environmental Protection, BSA
Robert Gordon, Project Manager, Contracting Officer, DOE/SC-BHSO
Timothy Green, Natural and Cultural Resources Manager, EPD, BSA

Jennifer Higbie, Biologist, EPD, BSA
Balwan Hooda, Subject Matter Expert for Radiological Emissions, BSA
Robert Lee, Deputy Division Manager, EPD, BSA
Douglas Paquette, Subject Matter Expert for Groundwater, BSA
Caroline Polanish, NEPA Compliance Officer, DOE/SC-BHSO
Mark Toscano, Project Manager, BSA
Jeffrey Williams, Subject Matter Expert for Non-Radiological Air Emissions, BSA

7.0. List of Agencies Contacted and Presentations to Stakeholders

7.1. Agencies Contacted

DOE NEPA regulations, found in 10 CFR 1021.301, require that the host state be provided the opportunity to review and comment on the EA document prior to DOE's approval of the EA.

Copies of the draft EA were distributed to the following New York State offices:

New York State Governor's Office – Albany, NY

New York State Department of Environmental Conservation – Stony Brook, NY

Additional copies of the draft EA were also sent to the following agencies for information only:

Town of Brookhaven Supervisor's Office – Farmingville, NY

Central Pine Barrens Joint Planning and Policy Commission – Great River, NY

Congressman Tim Bishop's Office - Coram, NY

Long Island Regional Planning Board – Hauppauge, NY

Suffolk County Department of Health Services – Yaphank, NY

Suffolk County Executive's Office – Hauppauge, NY

LIPA - Uniondale, NY

7.2. Stakeholder Presentations

Presentations related to planning and development of the proposed BP Solar Project were provided to the following stakeholder groups:

Brookhaven Executive Roundtable (BER)

The BER is a forum for frequent, routine and executive-level communications about BNL.

Represented on the BER are the major stakeholders associated with BNL, including the owner, operator, and jurisdictional, regulatory, oversight, community and political interests. Presentations about the BP Solar Project were provided to the BER on April 22, and September 23, 2009.

BNL Community Advisory Council (CAC)

The CAC consists of approximately 27-member organizations representing business, civic, education, employee, environment and health organizations.

Members meet monthly, set their own agenda, and work to reach consensus recommendations on issues of concern to them. Meetings are open to the public; each meeting has a comment period during which community members may voice their opinions and concerns [http://www.bnl.gov /community/CAC.asp].

Presentations about the BP Solar Project were provided to the CAC on April 15, and September 10, 2009.

The Central Pine Barrens Joint Planning and Policy Commission was contacted and a presentation was provided on September 16, 2009.

References

[Article 12]
Suffolk County Sanitary Code. (December 16, 1992). Article 12: Toxic and Hazardous Materials, Storage and Handling Controls. Suffolk County Department of Health Services, Code of Administrative Regulations Concerning Toxic and Hazardous Materials, Storage and Handling Controls.

[Bernstein 2009a]

Bernstein, D., et. al. (2009). A Stage I Archeological Survey for the Proposed Solar Array - Brookhaven National Laboratory. The Institute for Long Island Archeology, Department of Anthropology, State University of New York at Stony Brook, NY. August 2009.

[Bernstein 2009b]

Bernstein, D. & Merwin, D. (2009). Archeological and Architectural Data Recovery for the Privy Site at Brookhaven National Laboratory. The Institute for Long Island Archeology, Department of Anthropology, State University of New York at Stony Brook, NY. August 2009.

[BNL 2003a]

Brookhaven National Laboratory. (2003). Natural Resource Management Plan for Brookhaven National Laboratory. Brookhaven National Laboratory, Upton, NY. BNL-71870.

[BNL 2003b]

BNL. (2003b). *Wildland Fire Management Plan for Brookhaven National Laboratory*. BNL-71629-2003. Brookhaven National Laboratory, Upton, NY.

[BNL 2004]

Brookhaven National Laboratory. (2003). BNL Master Plan Update. August 2004. Brookhaven National Laboratory, Upton, NY.

[BNL 2005]

Brookhaven National Laboratory. (2005). Cultural Resource Management Plan for Brookhaven National Laboratory. Brookhaven National Laboratory, Upton, NY. BNL Report No. 73839.

[BNL 2009a]

Brookhaven National Laboratory. (2009). Land Use Controls Management Plan. Revision 3 (Final). Brookhaven National Laboratory, Upton, NY. June 10, 2009. https://luic.bnl.gov/website/landcontrols/documents/LUCMP_Final _2009.pdf

[BNL 2009b]

Brookhaven National Laboratory. Factsheet: Former Landfill Area (AOCs 2A, 2D & 2E). February 2009. https://luic.bnl.gov/website/landcontrols/factsheet/factsheet.cfm?factsheetID=1530

[Caithness]

Caithness Long Island, LLC. Long Island Power Authority Caithness Long Island Energy Center Final Environmental Impact Statement. June 2005.

[CERL]

Construction Engineering Research Laboratory (CERL). (1978). Construction-Site Noise Control, Cost-Benefit Estimation Technical Background. U.S. Army. January 1978.

[CPB]

Central Pine Barrens Joint Planning and Policy Commission. (1995). Central Pine Barrens Comprehensive Land Use Plan, Volume 1: Policies, Programs and Standards: as amended Oct. 2004.

[Davis]

Davis, Mark. (July 2009). Brookhaven National Laboratory Memorandum M. Davis to T. Green, *Estimated Population Residing Within 1-mile of BNL Site Boundary*, Brookhaven National Laboratory.

[DOE 1999]

U.S. Department of Energy Office of Environmental Policy & Assistance, RCRA/CERCLA Division (EH-413) *EH-413 Regulatory Bulletin - Hazardous Waste Identification Rule* Hazardous Remediation Waste Management Requirements (HWIR-Media): Final Rule *Effective Date: June 1, 1999*

[DOE 2009]

Climate Change Considerations in Project Level NEPA Analysis. January 13, 2009.

[DOE/EA-1662]

DOE/EA-1662 Environmental Assessment for 10 CFR Part 430 Energy Conservation Program: Energy Conservation Standards for Certain Consumer Products (Dishwashers, Dehumidifiers, Microwave Ovens, and

Electric and Gas Kitchen Ranges and Ovens) and for Certain Commercial and Industrial Equipment (Commercial Clothes Washers) April 2009

[eGrid]
eGrid 2007 version 1.1., THE EMISSIONS & GENERATION RESOURCE INTEGRATED DATABASE FOR 2007 - Technical Support Document for Year 2007; September 2008. http://www.epa.gov/cleanenergy/documents/egridzips/eGRID2007TechnicalSupportDocument.pdf

[Hailey]
Hailey, Solomon G, et.al. Greenhouse Gas Mitigation in Forest and Agricultural Lands: Reducing Emissions. Publication No. AE443, Agricultural and Biological Engineering Department, Florida Cooperative Extension Service, Institute of Food and Agricultural Sciences, University of Florida. Gainesville, Florida.

[EPA 1974]
U.S. Environmental Protection Agency (EPA). 1974. Information Levels of Environmental Noise Requisite to Protect Public Health and Welfare with an Adequate Margin of Safety. USEPA/ONAC 550/9-74-004. March 1974.

[EPA 2003]
U.S. Environmental Protection Agency (EPA). (April 2003).
http://www.epa.gov/sequestration/faq.html
http://yosemite.epa.gov/OAR/globalwarming.nsf/UniqueKeyLookup/LHOD5MJTBB/$File/2003- final-inventory_annex_o.pdf

[EPA 2006]
U.S. Environmental Protection Agency. Carbon Sequestration in Agriculture and Forestry. (October 2006). URL: http://www.epa.gov/sequestration/faq.html

[ERM]
Environmental Resources Management Southwest, Inc. Austin, Texas Noise Analysis for Upton Solar Project July 17, 2009 Memorandum Ponebshek, ERM SW to Richard Chandler, BP Solar July 17, 2009

[FEMA]

Federal Emergency Management Agency Flood Insurance Rate Map, for Suffolk County, NY, #36103C0730 G.

[HFBR EIS]

Draft Environmental Impact Statement Report for the High Flux Beam Reactor at the Brookhaven National Laboratory, Upton, NY. November 1999.

[Hoey]

Hoey, Steve. (April 1994). Brookhaven National Laboratory Natural Phenomena Hazards Evaluation. An attachment to the BNL Implementation Plan as per DOE Accelerator Order 5480.25.

[Kamer]

Kamer, Pearl M. (October 2005). The Economic Impact of Brookhaven National Laboratory on the NY State Economy. Brookhaven National Laboratory, Upton, NY.

[LIPA]

Long Island Power Authority. Population Survey 2007. (November 2007).

[LIRR]

Long Island Railroad. Official Timetables, May 18 - September 7, 2009. Greenport Service Timetable. http://mta.info/lirr/html/ttn/riverhea.htm http://mta.info/lirr/html/ttn/riverhea.htm

[Merwin]

Merwin, D. (2003). Archeological Field Survey Requirements for the Brookhaven National Laboratory. The Institute for Long Island Archeology, Department of Anthropology, State University of New York at Stony Brook, NY. October 2003.

[Nagle, 1975]

Nagle, C. M. (November 1975). Climatology of Brookhaven National Laboratory, 1949-1973. *BNL Report* No. 50466

[Nagle, 1978]
Nagle, C. M. (May 1978). Climatology of Brookhaven National Laboratory: 1974-1977. *BNL Report* No. 50857

[NYSDEC]
New York State Department of Environmental Conservation. Tiger salamander breeding pond protocol. 8pp NYSDEC, 1994.

[SC]
Suffolk County. Census 2000 Demographic Profile. 2009 http://www.co.suffolk.ny.us/Home/departments/planning/Census 2000 Demographic Profiles.aspx - br

[SER]
2007 Site Environmental Report. (October 2008). Brookhaven National Laboratory, Brookhaven Science Associates, P.O. Box 5000, Upton, NY 11973-5000. BNL Report No. 81352-2008.

[Tele Atlas]
Tele Atlas North America, Inc., and ESRI, 20080401, U.S. Census Block Groups: ESRI® Data & Maps 2008 World, Europe, United States, Canada, and Mexico, ESRI, Redlands, California, USA.

[U.S. Census, 1999]
U.S. Census Bureau. Poverty Thresholds 1999. U.S. Census Bureau, Housing and Household Economic Statistics Division. August 2006. http://www.census.gov/hhes/www/poverty/threshld/thresh99.html

[U.S. Census, 2008]
U.S. Census Bureau. Poverty Thresholds 2008. U.S. Census Bureau, Housing and Household Economic Statistics Division. February 2008. http://www.census.gov/hhes/www/poverty/threshld/thresh08.html

[USGS. 1998]
National Earthquake Information Center, U.S. Geological Survey Earthquake Data Base, Search of Historical and Primary Data 1973-1997, U.S. Geological Survey.

End Notes

[1] *Brownfields* are defined as abandoned, idled, or underused industrial and commercial facilities where expansion or redevelopment is complicated by real or perceived environmental contamination [DOE 1999]

[2] NOTE: The Laboratory's on-site population is not shown on Figure 6.

INDEX

C

D

F

G

H

I

J

L

M

N

O

P

R

S

T

U

V

W

Y